Handbook for
Developing and Managing
Tribal Nonpoint Source Pollution Programs
Under Section 319 of the Clean Water Act

Foreword

Congress amended the Clean Water Act (CWA) in 1987 to establish the section 319 Nonpoint Source Management Program in recognition of the need for greater federal leadership to help focus state, tribal, and local nonpoint source efforts. Under section 319, states, territories, and Indian tribes receive grant money that supports a wide variety of activities including technical assistance, financial assistance, education, training, technology transfer, demonstration projects, and monitoring to assess the success of implementing management practices that address pollution from nonpoint sources.

As of the time of publication of this document, 159 tribes have approved nonpoint source programs. Tribal water quality programs continue to increase in number and to mature in their capacity to understand and improve the condition of reservation lakes, rivers, streams, wetlands, and coastal waters. In support of the continued growth and sophistication of tribal participation in the CWA section 319 program, the U.S. Environmental Protection Agency (EPA) is pleased to release this *Handbook for Developing and Managing Tribal Nonpoint Source Pollution Programs Under Section 319 of the Clean Water Act.*

EPA is committed to restoring and protecting our waters through a watershed approach, and it is encouraging to see a number of tribes electing to pursue funding to develop watershed-based plans. Cooperative, on-the-ground, watershed-based efforts among tribal and nontribal water resource managers and staff are helping to improve the prospects for solving water quality problems that know no boundaries, and affect the health and quality of life of all Americans.

This handbook is meant to be a practical and accessible guide for tribes to answer key questions such as

- How do I develop a nonpoint source assessment report and management program that meet 319 program eligibility requirements and set the stage for effective program implementation?

- What sorts of activities are eligible for funding under CWA section 319?

- How do I develop and successfully implement a watershed project that will help restore the quality of our water for drinking, fishing, and other uses?

The handbook explains the role of both EPA and the tribes in working together to help solve water quality problems caused by nonpoint source pollution. All aspects of the grants-funding process are broken down for you in simple steps, showing you how tribes can use section 319 program funds to implement programs and projects to reduce pollution and restore water quality. At the same time, it takes you the next level by providing a great deal of useful technical information regarding nonpoint source pollution, how you can develop and assess available data to develop a plan of action, and what management practices and activities are needed to solve the problem.

EPA is proud of the many excellent projects that tribes have already implemented with section 319 funds. Many impressive tribal on-the-ground projects are available for you to review on *www.epa.gov/nps/tribal*. It is our hope that tribes will find this handbook a useful tool in helping to achieve many more water quality successes in the future.

Benita Best-Wong
Director of the Assessment and Watershed Protection Division
Office of Water

Contents

Figures

Tables

Introduction

There is a deep spiritual connection between Native American people and the earth. Tribal communities are strongly committed to the restoration and protection of the natural environment, including surface and ground water resources. These rivers, lakes, streams, reservoirs, wetlands, estuaries, and coastal waters sustain fish and shellfish, provide recreational opportunities, supply drinking water, and allow ceremonial uses for many tribal communities. However, many water resources are threatened or impaired by polluted runoff, also known as nonpoint source (NPS) pollution. The goal of this handbook is to provide tribes with guidance and other information that will help them to protect and restore water resources.

What this Document Contains

This edition is an update of the *Tribal Nonpoint Source Planning Handbook* (EPA document no. EPA-841-B-97-004), which was issued in September 1997. It contains new information and additional features. The handbook has three parts. **Part I** describes the Clean Water Act (CWA) section 319(h) grant application process and explains in detail how tribes can demonstrate their eligibility and prepare the necessary documentation for successful 319 grant applications. New tribal examples showcase useful approaches for protecting or restoring water quality, and Internet resources provide tools for tribes interested in growing their NPS programs. **Part II** provides detailed information about the watershed-based approach to solving NPS problems, as well as ideas for leveraging funds. Additional resources for tribes are provided in **Part III**, which offers useful links to various resources, online management tools, and a list of contacts.

Key Questions

What Is NPS Pollution?

NPS pollution—polluted runoff—occurs when rainfall, snowmelt, or irrigation water runs over land or through the ground, picks up pollutants, and transports them into surface waters or ground water. Though the relative impact from a few nonpoint pollution sources might be small, the cumulative effect from many nonpoint sources degrades water quality. In fact, NPS pollution is the leading source of water quality problems in the United States. Major nonpoint sources of pollution often include agricultural practices; unrestricted livestock grazing; poor siting and design of roads, highways, and bridges; forestry; urban runoff; abandoned mines; construction sites; channelization of streams; and hydromodification, such as building and maintaining dams and levees. Other sources include lawn and garden maintenance, malfunctioning septic systems, constructing marinas, boating, and storm drain dumping.

Atmospheric deposition of pollutants originating from power plants, factories, trucks, and automobiles is also considered NPS pollution.

What Is the Status of Our Nation's Waters?

The Wadeable Streams Assessment, a national survey of streams conducted by EPA and the states and published in December 2006, found that 42 percent of U.S. stream miles are in poor biological condition, 25 percent are in fair condition, and 28 percent are in good condition. The most widespread stressors observed across the country and in each major region are nitrogen, phosphorus, riparian disturbance and clearing, and streambed sediments. These stressors are often associated with polluted runoff. The survey also found that 30 percent of the nation's streams have elevated levels of nitrogen and phosphorus, and 25 percent have elevated levels of sediment. For more information, see the *Wadeable Streams Assessment* at *www.epa.gov/owow/streamsurvey*.

In addition, the latest published summary of state water quality assessments found that agricultural activities were the leading source of impairment in assessed rivers and streams, associated with more than 94,000 impaired stream miles. Other leading nonpoint pollution sources in streams include hydromodification (e.g., streambank destabilization and flow alteration), riparian habitat alteration, and urban runoff. For more information, see the *National Water Quality Inventory: Report to Congress, 2004 Reporting Cycle*, January 2009, at *www.epa.gov/owow/305b/2004report*.

In assessed lakes, ponds, and reservoirs, agricultural activities ranked near the top of the identified pollution sources, associated with more than 1.6 million impaired lake acres. Hydromodification was associated with more than 1.2 million impaired lake acres; urban runoff was associated with 701,000 impaired lake acres; and unspecified nonpoint sources affected nearly 500,000 impaired lake acres. For the full report, see the *National Water Quality Inventory: Report to Congress, 2004 Reporting Cycle* at *www.epa.gov/owow/305b/2004report*. Note that only about 16 percent of U.S. streams miles and 39 percent of lake acres were assessed for this report; some states did not report on sources of impairment in their waters, and more than one source might impair any given waterbody.

What Are Management Measures or Best Management Practices?

NPS pollution can be addressed using various *management measures*, which are the best available economically achievable practices needed to solve a water quality problem. Another common term used is *best management practices* (BMPs). A BMP can be a particular technique, measure, or structural control used to manage the quantity and improve the quality of runoff to the maximum extent possible and most cost-effectively. The U. S. Department of Agriculture (USDA) and others have identified a number of BMPs to address pollution from different sources. Such management practices vary widely, and they must be tailored to specific conditions. For example, BMPs to protect waters from agricultural pollution, such as sediment from row crop land, are different from urban runoff BMPs, such as those that address sediment from construction sites. BMPs can be structural, such as fences to prevent livestock from entering a stream, or nonstructural, such as an ordinance that specifies vegetated buffer zones between certain activities and the water's edge. For more information on BMPs, visit EPA's Web site at *www.epa.gov/nps* and click on *NPS Categories*. Each NPS category will lead you to a series of guidance documents for your category of interest. For even more information, visit the Natural Resources Conservation Service (NRCS) Web site at *www.nrcs.usda.gov/technical/efotg*. Another excellent resource to view various BMPs and learn what they do has been compiled in an appendix available at *www.waterquality.utah.gov/TMDL/Virgin_River_Watershed_Implementation_Appendix.pdf*.

What Is the Watershed Approach?

EPA and other entities both inside and outside the government believe that the watershed approach is the most promising means of addressing NPS pollution and the challenging condition of our water resources. The watershed approach focuses efforts on a particular *watershed*, which is the area of land that drains to a specific point, such as the confluence of two rivers, a lake, or a coastal estuary (Figure i-1). The watershed approach is characterized by these unique features:

1. It is hydrologically defined.

   ■ Has a geographic focus based on hydrologic connections (i.e., a watershed or drainage area)

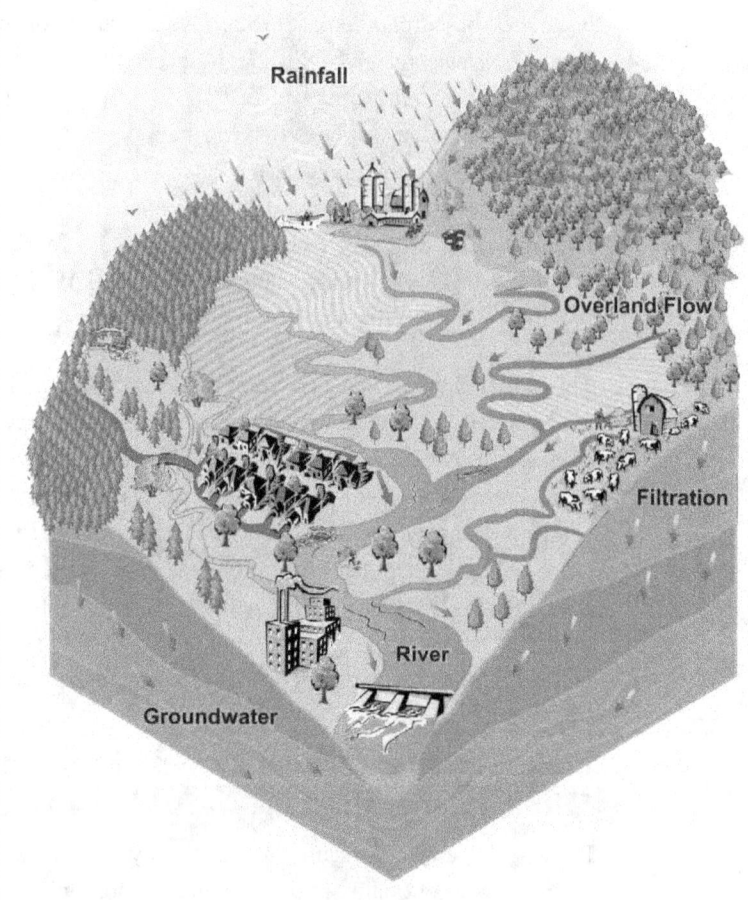

Figure i-1.   Illustration of a watershed.

  ▣ Includes all stressors/causes and pollutant sources within that geographic area

2. It involves all stakeholders.

  ▣ Includes public (federal, tribal, state, local), private, and nongovernmental sectors

  ▣ Is community-based

  ▣ Includes a coordinating framework

3. It strategically addresses priority water resource goals.

  ▣ Integrates multiple regulatory and voluntary programs

  ▣ Is based on sound science

  ▣ Uses adaptive management for continual improvement

The watershed approach provides a framework for working on a watershed basis and is used to generate a *watershed-based plan that addresses impairments and threats to water quality*. States are already required to show that they are working toward achieving elements of a watershed-based plan in order to receive a certain portion of their CWA section 319 funding (i.e., the *incremental funds*). Although developing and implementing watershed-based plans are not requirements for tribal section 319 funding as of the date of this publication, EPA strongly encourages tribes to develop such plans to comprehensively understand and plan for protection and restoration of water quality. EPA believes that using the watershed approach is usually more effective than addressing impaired waters piecemeal, such as one segment of a river or stream reach at a time. In addition to 319 funding, tribes can explore the use of other funding, such as General Assistance Program (GAP) funding and CWA section 106 (Water Pollution Control Program) funding, to support the development of watershed-based plans.

This does not mean that previous assessment work that has been done is obsolete. The information you have compiled in your assessment report and management plan (to be discussed later), or in any other previously developed planning document, will feed easily into a more comprehensive watershed plan. For example, the watershed-based plan might span a 5-year time frame and include all lands that are part of a tribal watershed, such as surrounding federal, state, and private lands and their associated pollutant sources and stressors. To effectively be this inclusive, tribes will need to work with all the stakeholders in their watersheds, both tribal and nontribal. This might seem tricky at first, but these efforts will result in significant progress toward reaching your water quality goals.

How Can I Integrate My NPS Program with Other Water Quality Programs?

Because the federal government has steadily moved toward a watershed approach for addressing water quality protection and restoration issues, there is now a greater need to communicate across various environmental programs. EPA programs include water quality standards, CWA section 106, CWA section 303(d), National Pollutant Discharge Elimination System (NPDES) wastewater discharge permitting, monitoring, source water protection, and solid waste management. Figure i-2 shows how these water programs relate to and support each other. EPA encourages tribal NPS program staff to interact with their counterparts in other programs to coordinate activities and leverage funding resources to reach common water quality goals. Part II of this handbook also provides information on some of these programs.

Figure i-2. Water quality program integration.

History of the CWA Section 319 Program

In 1987 Congress amended the CWA to add sections 319 and 518. That provided support for states, territories, and tribes to address NPS pollution. NPS pollution is increasingly recognized as the most pervasive source of water quality impairments in the nation, far outweighing problems from wastewater treatment plant, industrial facilities, stormwater runoff, and other discharges in most areas.

CWA section 319 authorizes EPA to award grants to eligible tribes, states, and intertribal consortia to implement EPA-approved NPS management programs developed pursuant to section 319(b). The primary goal of an NPS management program is to control or prevent NPS pollution. This is accomplished by implementing BMPs that reduce pollutant loadings to waterbodies from each NPS category or subcategory identified in the tribe's NPS assessment report. The NPS assessment report is developed pursuant to section 319(a).

CWA section 518 authorizes EPA to treat federally recognized Indian tribes in a similar manner as states ("treatment as a state," or TAS). EPA recognizes that TAS does not necessarily capture the true relationship between tribes and the federal government. For example, EPA Region 9 uses the term *Financial Assistance Eligibility (FAE)* instead of TAS to describe eligibility status. For the purposes of this handbook, TAS will be used, given its prominence in statute and regulation.

CWA section 518(f) states that no more than one-third of one percent of the total amount of section 319 funds appropriated for any fiscal year may be used to make grants to tribes. EPA recognizes, however, that this statutory cap would most likely substantially affect tribes' ability to establish and maintain NPS programs. There is an understanding that Indian tribes need more financial support to implement NPS programs. In addition, more tribes are receiving TAS status under the 319 program and receiving funds. For those reasons, since 2000 Congress has approved EPA's annual requests to exceed the statutory cap. For a complete list of section 319-eligible tribes, see the document listed under *Funding* at *www.epa.gov/nps/tribal*.

PART I

The Clean Water Act Section 319 Program

Section 319 Program at a Glance

What it does: Requires tribes and states that wish to receive section 319 funding to prepare an assessment of their NPS pollution problems and develop a management program to address the problems identified in their assessment report. Also creates a grant program for states and tribes to implement their approved programs, including implementation of on-the-ground projects to reduce NPS problems.

What it funds: NPS program staff, outreach and education activities, travel and training associated with NPS activities, NPS ordinance development, and various on-the-ground BMPs. Proposed activities must implement management measures identified in the management plan.

Who is eligible: Tribes and states are eligible to receive direct funding from EPA through congressional appropriations. Tribes must have TAS and an approved NPS assessment report and NPS management program to receive 319 funds. Tribes and states must make *satisfactory progress* to continue receiving 319 funds every year, as well as develop approved work plans. Although eligibility varies from state to state, tribes (along with other entities) are also eligible to apply to the state to receive 319 funding.

How it is funded: Congress appropriates funds every year for the section 319 program. In the past few years (as of 2010), this appropriation has been approximately $200 million on an annual basis, but the amount varies from year to year (to view annual appropriations, visit *www.epa.gov/nps/319hhistory.html*). Tribes have a statutory limit of receiving only one-third of one percent of that appropriation; however, since 2000 Congress has annually allowed EPA to exceed this statutory cap. States receive base and incremental funds (i.e., funds used to develop and implement watershed-based plans that address NPS impairments in watersheds that contain 303(d)-listed waters) to support their NPS management program. Tribes receive base funds that support their NPS management program, and they are eligible to compete nationally for additional funds to implement projects to restore and protect their waters.

Activities Eligible for Funding under CWA Section 319

Activities eligible for funding under section 319 are quite extensive and cannot be listed in their entirety, but the following list illustrates some of the activities funded under this program. A tribe must propose activities that are discussed in its EPA-approved NPS management plan, which is described further in the section titled Nonpoint Source Management Program Plan for 319(h) Eligibility, which begins on page I-44.

- NPS training for tribal staff
- Developing NPS education programs
- Hiring an NPS coordinator
- Developing watershed-based plans
- Road stabilization/removal
- Riparian planting
- Stream channel reconstruction
- Wetland development for sediment/toxic substance removal
- Low-impact development projects/stormwater mitigation
- Riparian livestock exclusion fencing/off-site watering
- Springs protection
- Outhouse removal/rehabilitation
- Large woody debris placement
- Lake protection and restoration activities
- Ground water activities
- Urban stormwater activities; remediation of abandoned mine lands; or activities related to animal waste storage, treatment, and disposal not specifically required by NPDES permits
- Invasive/nonnative species removal (must demonstrate link to water quality impairment)
- Project monitoring (pre-project, post-project)
- NPS ordinance development

For more examples of activities eligible under section 319, see EPA's NPS *success stories* at *www.epa.gov/nps/success*. That Web site features stories about restoration efforts that have led to documented water quality improvement of NPS-impaired waterbodies.

Activities Ineligible for Funding under CWA Section 319

Although a multitude of NPS activities may be funded through the 319 program, other NPS-related activities are typically ineligible within the context of the grant. For example, some of these prohibitions are based on statutory limitations on the use of 319 funding.

If you have questions regarding any of these activities, contact the EPA tribal NPS coordinator in your region. In addition, you can find information on the following ineligible activities in the most recent 319 guidelines at *www.epa.gov/fedrgstr/EPA-WATER/2003/October/Day-23/w26755.htm*.

- General monitoring unrelated to assessing the control of nonpoint sources of pollution

- Implementing specific requirements of NPDES point source or stormwater permits

- Implementing the permit application requirements of EPA's stormwater regulations (e.g., mapping stormwater systems, identifying illicit connections, characterizing stormwater discharges, or monitoring or BMP treatment required by an NPDES permit)

- Operation and maintenance of NPS implementation projects

- NPS activities not identified in your NPS management program

Summary of Eligibility Requirements for an NPS Grant

Four eligibility conditions must be met before tribes or intertribal consortia may submit a work plan for funding under an NPS (section 319(h)) grant. (Work plans are discussed in detail in the section titled Work Plan Development, which begins on page I-68.) To receive funding, the tribe must do the following:

1. Be *federally recognized*

2. Have an *approved NPS assessment report* in accordance with CWA section 319(a)

3. Have an *approved NPS management program* in accordance with CWA section 319(b)

4. Be *approved for treatment in a similar manner as a state* (TAS) in accordance with CWA section 518(e)

The flowchart in Figure I-1 describes eligibility requirements, eligible activities for the two levels of 319 funding (base and competitive, which are explained in detail in the section titled Section 319 Funding Process, which begins on page I-62), and the general grant process.

The treatment as a state (TAS) section of this handbook, which begins on page I-5, discusses section 518 of the CWA, which authorizes EPA to treat federally recognized Indian tribes in the same manner as states. It also contains an overview of what is needed for TAS status and provides information on the process of applying for TAS. Because federal recognition is an integral part of the TAS process, those topics are discussed together below.

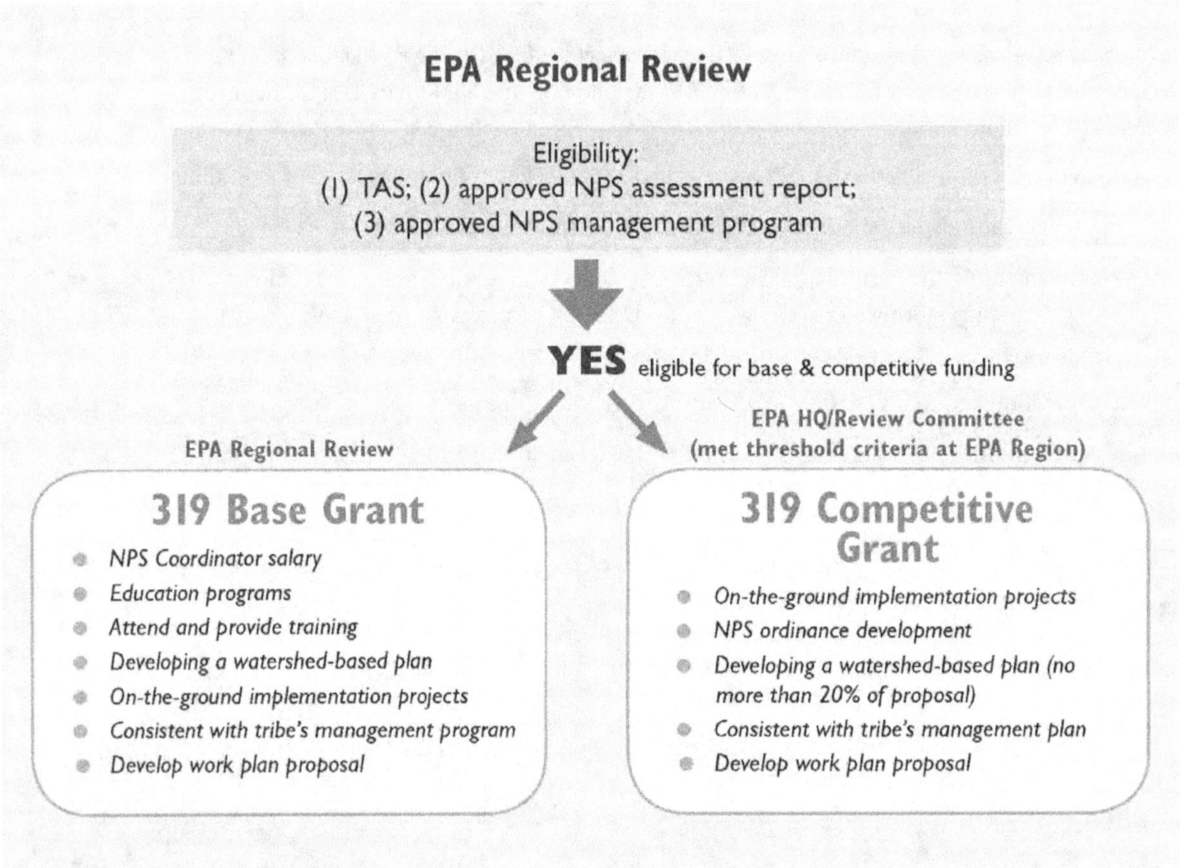

EPA Regional Review

Eligibility:
(1) TAS; (2) approved NPS assessment report;
(3) approved NPS management program

YES eligible for base & competitive funding

EPA Regional Review

EPA HQ/Review Committee
(met threshold criteria at EPA Region)

319 Base Grant

- NPS Coordinator salary
- Education programs
- Attend and provide training
- Developing a watershed-based plan
- On-the-ground implementation projects
- Consistent with tribe's management program
- Develop work plan proposal

319 Competitive Grant

- On-the-ground implementation projects
- NPS ordinance development
- Developing a watershed-based plan (no more than 20% of proposal)
- Consistent with tribe's management plan
- Develop work plan proposal

Figure I-1. Eligibility and application requirements for base vs. competitive grants.

Detailed Guide to the 319(h) Grant Eligibility Components

Federal Recognition and Treatment in the Same Manner as a State for 319(h) Funding Eligibility

For tribes to receive *treatment as a state* (TAS) authorization for implementing certain CWA programs, they must demonstrate that

1. They are federally recognized.

2. They have a governing body capable of carrying out substantial governmental duties and powers.

3. The functions they seek to exercise pertain to the management and protection of water resources on tribal lands.

4. They are reasonably expected to be capable of carrying out the functions in a manner consistent with the CWA and its regulations.

Statutory and regulatory text regarding these requirements can be found in section 518(e) of the act and at Title 40 of the *Code of Federal Regulations* (CFR) sections 35.633 and 130.6(d).

The following sections outline examples of the documentation tribes may use to show that they meet these requirements. Tribes submitting information to EPA should clearly label it and attach to the front of the package a list of the attachments, exhibits, or supporting documents.

> In this section, you will find
> • *The documentation tribes need to provide to become federally recognized and obtain treatment in a similar manner to a state (TAS)*

Taos Pueblo, New Mexico

Benefits of restoration, Rio Pueblo.

Federal Recognition

Federally recognized tribes are those on the list of *Indian Entities Recognized and Eligible to Receive Services from the United States Bureau of Indian Affairs*, which is published in the *Federal Register* by the Department of the Interior. In 2009 a total of 564 tribes were on the list. To meet the requirements for federal recognition, tribes should provide a *Federal Register* citation to their presence on the current list.

Substantial Governmental Duties and Powers

Tribes must demonstrate that they carry out substantial governmental duties and powers to administer and implement an NPS program. Some examples of information a tribe can submit follow:

- A description of the form of tribal government, including its executive, legislative, and judicial powers

- A list of functions or programs the tribal government currently performs

- A description of the source of tribal government authority, such as the tribal constitution, by-laws, or ordinances

The activities that tribes can describe include those that promote the public health, safety, and welfare of the people, such as acquiring land, adopting and implementing ordinances, exercising police power, and so on. Descriptions of governmental functions or programs can include Web site references for key documents, such as the tribal constitution and ordinances. EPA prefers summary descriptions of this information over copies of the referenced documents, and a tribal organizational chart is helpful.

Reservation Waters

Tribes should provide a certification statement from the tribal attorney demonstrating that the activities they will carry out pertain to the management and protection of reservation water resources (which includes water resources of tribal trust lands even if those lands have not been designated formally as reservations). Such information should include maps or legal descriptions of the reservation areas (or both), descriptions of the water resources of the reservation areas, and descriptions of how the activities will pertain to management and protection of those water resources. That statement should also cite the treaty or other history of the tribe establishing the reservation and its current boundaries, or any other authority by which the tribe's trust lands were established.

Tribal Capability

Tribes must have the capability to manage a section 319 program. To demonstrate capability, tribes should include information supporting their existing ability to carry out functions such as planning; identifying staff and other resources; budgeting; developing work plans and schedules; carrying out the identified work tasks; assessing their impact or results; and handling financial disbursements, reports, and other actions through a system with appropriate controls. Tribes may demonstrate their capability to manage and implement an NPS pollution control grant by describing the organization that will carry out the program, any existing or current environmental programs being carried out, mechanisms in place for carrying out the governmental functions of the tribe, and the technical and administrative capabilities of the staff that will manage the program.

Tribal Management Experience

Supporting documentation may include brief descriptions of federal, tribal, and other programs managed by the tribe, including those involving natural resources, environmental issues, drinking water, or others. This documentation may overlap somewhat with tribal responses regarding capability, existing programs, and technical/administrative capabilities of the staff. Examples for addressing this requirement include quarter-page, half-page, or page-long summaries of programs currently managed by the tribe, a list of programs managed by the tribe over the past three years, or a summary report of tribal operations that conveys a sense of the tribe's management experience.

Existing Programs

In conjunction with the requirement, the tribe may provide a list of current programs managed by the tribe. The existing programs list may include supplemental information, such as number of staff, total budget, and so on. Programs to list include health clinics and related programs, Bureau of Indian Affairs programs, fish and game programs, agricultural and land management programs, tribal administrative office functions, and others.

Mechanisms for Governmental Functions

Supporting documentation may include a summary of tribal operating procedures, tribal constitutions or by-laws, or other documents that describe how decisions are made, how rules or ordinances are established, how governmental affairs are handled, and the like. The tribe should clearly and succinctly describe how its government functions, with references to charters, constitutions, by-laws, or other documents that establish the mechanisms for governmental functions. If that information is already provided in the description of governmental functions portion of the application, the information may be referenced, without repeating it, in this section.

Accounting and Procurement Systems

The tribe may include descriptions of its accounting and procurement systems, sections from the tribal charter or by-laws or other documents describing the tribe's financial system, copies of the tribal financial operating procedures, or similar documents that demonstrate that the tribe has a workable system in place, with adequate controls to prevent poor accounting practices or misconduct. The summary should be brief and should clearly lay out how the accounting and procurement systems function.

▓ *Technical and Administrative Capabilities of the Staff*

This section should include staff resumes, number of years of relevant experience, brief summaries of the technical and administrative experience and qualifications of staff that will be involved in the NPS pollution management program, or personal descriptions of qualifications and experience that convey an ability to manage the program. Include summaries for nontribal technical assistance and other personnel if they will be involved in the tribe's NPS pollution management program.

Generally those documents can also be found in the CWA 106 applications, so there is no need to duplicate efforts if you already have TAS for section 106. However, many of the CWA 106 applications are several years old, and updated forms might be appropriate. You may develop TAS documentation independently or in parallel with the NPS assessment report and management program, also required for 319(h) eligibility (discussed below). Each EPA Regional office has an approval process and time frames within which these critical documents are processed. For assistance, contact your Regional Tribal NPS coordinator.

Nonpoint Source Assessment Report for 319(h) Eligibility

What Is an Assessment Report?

The assessment report is a comprehensive technical summary of the condition of tribal water resources. The report provides the foundation for the scope and direction of the NPS pollution management program, discussed in the next section. It is important to characterize all waterbodies, including, if possible, those that might lie partly on nontribal lands, making partnerships with external parties important to the overall process. Two types of information that will determine the scope and direction are:

> In this section, you will find
> - *A detailed explanation of the next eligibility requirement— an approved NPS assessment report*
> - *A suggested format for the assessment report*
> - *Excerpts from successful tribal assessment reports*
> - *References and resources to help you write your assessment report*

1. Actual NPS-related *impairments*, to target restoration efforts

2. Waterbodies of high quality or cultural significance that need *protection* from existing or future sources of polluted runoff

The NPS assessment report must include four types of information (which are also referred to as the four *legislative conditions*):

1. An identification of *waters that cannot be expected to attain or maintain tribal water quality standards* without the control of NPS pollution. If the tribe does not have water quality standards, state standards or tribal water quality goals may be used as guides for evaluation of water quality. This is a place to use your 106/303/GAP data and information and any others sources of data that meet established quality assurance/ quality control (QA/QC) criteria.

2. An identification of the *categories and subcategories* of NPS pollution that contribute to the water quality problems for the individual waters identified in number 1 above. For a listing of major NPS pollution categories and subcategories, refer to the latest *Guidelines for the Preparation of State Water Quality Assessments (305(b) Reports)*, published by EPA. Tribes may also use the reference information provided by EPA's Web site at *www.epa.gov/nps/categories.html*.

   Look at the activities identified that impair or threaten water quality and link them back to these categories/subcategories. Many states have their 303(d) and 305(b) reports online, and those could provide an initial resource.

3. A description of *how the tribe will identify the BMPs* needed to control each category and subcategory of NPS pollution identified in number 2 above, as well as a description of how the management practices will be used to reduce the level of pollution resulting from these sources. Such factors as public participation and inter/intragovernmental coordination should be included. Many tribes use state or federal tools and resources, such as Internet databases or publications, to identify the kinds of BMPs that will address or prevent the NPS issues identified in their waters.

   The assessment report should include a description of how those sources were identified and how the specific practices were chosen. It will be important to carefully consider the pollutant sources, so that the appropriate management practices can be selected. An efficient and effective way to address this category of information is to establish a technical committee to review the BMPs already identified in the surrounding region. If the tribe has worked with NRCS and is using that agency's technical specifications, that information could be included.

4. *A description of any existing tribal, state, federal, and other programs* that might be used for controlling NPS pollution. This section should include federal,

The following is a sample list of subcategories for agricultural NPS pollution.

Agriculture
- *non-irrigated crop production*
- *irrigated crop production*
- *specialty crop production*
- *pastureland*
- *rangeland*
- *feedlots (all types)*
- *aquaculture*
- *animal holding/management areas*
- *manure lagoons*

state, tribal, local, and nonprofit programs that provide (or could provide) funding or technical support for NPS work on tribal lands. The tribe should describe programs that are in use, as well as those that are available to the tribe but have not yet been actively pursued. Doing a bit of research for this information might identify new programs that could provide partnerships and assistance for new NPS programs.

Although there might be many sources of water quality data, data used to develop the assessment report must meet QA/QC requirements or be selected through an approved secondary data analysis protocol. Data sources could include data collected using section 106 and other federal agency funds, as well as data collected by other agencies, such as the U.S. Geological Survey (USGS). Where monitoring data are not available, visual observations, such as the NRCS *Stream Visual Assessment Protocol* described at *www.nrcs.usda.gov/technical/ecs/aquatic/svapfnl.pdf*, may be used.

It is preferable for data to be geo-located and to show mean, median, ranges, and thresholds. The data need to be presented in a manner that documents impairments or threats based on water quality standards, narrative criteria, or water quality goals. The tribe's analysis of the data will be the guide for prioritizing the places and watersheds to work in, and the kinds of BMPs that will be needed to control NPS impacts or protect high-value waters.

Format of an NPS Assessment Report

Section 319(a) of the CWA specifies the information that must be included in tribal NPS assessment reports (the four types of information described above). To facilitate the preparation of these reports, a suggested outline for an NPS assessment report follows. The notice and opportunity for public review and comment are covered in the section entitled Public Notice and Comment, which begins on page I-60; they are requirements for both assessment and management reports.

- *Cover Page*—Title and the date (month and year) of the assessment
- *Table of Contents (but labeled simply as* Contents)—A listing of the major sections of assessment report, lists of figures and tables, appendices, and corresponding page numbers
- *Text (body of the report)*—According to the headings of each major section of the assessment report and corresponding page numbers
  - *Overview*
  - *Introduction*
  - *Methodology*
  - *Land Use Summary*
  - *Surface and Ground Water Quality*
  - *Results*

- *Discussion*
- *Selection of BMPs*
- *NPS Control Programs*

- *References/Sources of Information*—Some tribes might already have collected data that will help them develop an NPS assessment report. With that information, tribes can assess NPS pollution problems and determine baseline water quality data without completing additional water quality surveys. Tribes can also use data collected with CWA section 106 funds to help identify high-priority problems.

- *Appendices*

- *Acronyms and Abbreviations List*

The next section describes and provides examples for each component of an assessment report. Many of the examples that follow have been drawn from approved tribal NPS assessment reports. When developing and writing an assessment report, a tribe should take the time to make sure the report is well-documented, readable, and understandable by many different audiences, including tribal and EPA staff, tribal partners, and the public.

Overview

In the overview, state the purpose of the report and explain the need for an NPS assessment report for the tribal waters. The section should include a brief description of the reservation or other tribal lands, key NPS issues (why the report is being written, e.g., to establish baseline data for program, provide direction for the management program plan, or to meet some goal(s) that the tribe has for natural resources) and key conclusions. Also provide a general summary of the analysis that will follow, stressing major conclusions and broad areas of concern. Be sure to mention which categories and subcategories of NPS pollution were identified through your assessment. Discuss only significant data and general findings in this section. The section should be concise and ideally should not exceed one page. It is recommended that you write this section last so that it accurately reflects the rest of the report.

Example from the Confederated Tribes of Grand Ronde (CTGR 2008a)

This report has been created to assess nonpoint source water quality on or upstream of lands owned in Trust by the Confederated Tribes of Grand Ronde (hereafter, the Tribe) and provide background information for a nonpoint source management plan. The Tribe currently owns approximately 10,871 acres of land in Reservation or Trust Status, with an additional 129 acres pending conversion into Trust status (Table I-I).

Section 319 of the federal Clean Water Act provides authority to states, territories, and tribes to address problems associated with nonpoint sources of pollution. Furthermore, the U.S. Environmental Protection Agency uses section 319 as the primary source of funding to address nonpoint source problems. In order to qualify for section 319 grants, the Tribe must complete a nonpoint source assessment report and a nonpoint source management plan that are approved by the U.S. Environmental Protection Agency.

The goal of this assessment is to focus attention on water quality parameters and issues that point to significant or potentially significant nonpoint sources of pollution and provide guidance on how to monitor effectively and alleviate significant sources.

Table I-I. Size and status of land owned by Confederated Tribes of Grand Ronde (CTGR 2008a)

Map ID	Tract name	County	Acres	Status
1	Reservation	Yamhill	10,051.32	Reservation
2	Stimson	Yamhill	160.00	Trust
3	Risseeuw	Yamhill	78.14	Trust
4	Eastman	Yamhill	9.99	Trust
5	Natural Resources	Yamhill	36.61	Trust
6	King	Yamhill	11.93	Trust
7	AGZ	Yamhill	9.13	Pending Trust
8	Camping	Yamhill	8.90	Reservation
9	Elders' Housing	Yamhill	19.55	Reservation
10	Mahurin	Yamhill	10.26	Pending Trust
11	Smith	Yamhill	31.31	Pending Trust
12	Tribal Headquarters	Polk	118.92	Reservation
13	Procurement	Polk	2.00	Trust
14	Multi-family housing	Polk	19.73	Trust
15	Family housing 2	Polk	20.21	Trust
16	Cemetery	Polk	7.52	Reservation
17	Grand Meadows	Polk	10.76	Reservation
18	Windsor	Polk	7.90	Trust
19	I.P.	Polk	54.21	Pending Trust
20	Round Valley	Polk	24.33	Pending Trust
21	Old Depot	Polk	2.24	Reservation
22	Casino HR	Polk	4.69	Reservation
23	Casino	Polk	84.45	Reservation
24	Gould	Polk	6.68	Reservation
25	C-Store/RV Park	Polk	12.40	Reservation
26	North Busswell	Polk	58.47	Reservation
27	Fort Yamhill Park	Polk	113.53	Trust
28	New Pow-wow grounds	Polk	25.55	Trust

Example from the Shakopee Mdewakanton Sioux Community (SMSC 2008)

Fee and Trust lands of the Shakopee Mdewakanton Sioux Community (SMSC) consist of approximately 2,625 acres in Scott County, Minnesota. SMSC lands contain various types of surface waters including lakes, streams and wetlands, and three groundwater wells that provide drinking water for the community. Rapid urban development on SMSC lands impacts waters via nonpoint source pollution (NPS) such as nitrogen, phosphorous, salts, oil, heavy metals, and sediment. Potential sources of NPS pollution include agricultural runoff; commercial runoff from buildings and parking lots; urban runoff from roadways, turf areas and construction activities; and residential runoff from driveways and lawns. The principal goal of this pollution assessment is to evaluate current NPS pollution impacts on SMSC water and outline measures to ameliorate future impacts.

Surface water quality data, biological data, and groundwater well testing data with an ArcView Geographic Information System database were used to evaluate individual SMSC surface waters. NPS pollution has negatively impacted all surface waters located on SMSC lands. Impacts include degraded water quality, reduced fisheries and wildlife habitat potential and reduced aquatic species richness and diversity. Most SMSC waterbodies are moderately impacted, and provide partial support of primary and secondary contact, fisheries use, and full support for agricultural use based on State of Minnesota standards. However, one stream site located adjacent to the Reservation is seriously impacted by agricultural use, and does not support use categories for which it is classified. Proposed future land use changes as specified in the SMSC long-term comprehensive plan indicate more areas could become seriously impacted by NPS pollution. The SMSC adopted a Stormwater Management and Erosion Control ordinance in March 2003 to help protect water quality from development.

Introduction

The reader should come away from the introduction with an understanding of the reservation size, location, and fee/trust relationships and a summary of key water information—surface NPS pollution problems, the probable sources, and actions that can be taken to control them. The section should include a description of the tribe, including historical and land use development context and cultural issues. If the tribe has current land planning or other natural resources management programs, those programs should be described and related to the water quality assessment in this section. An example of that would be CWA section 106 goals.

The introduction should also cover the goals and objectives of the report and describe the public comment process. A goal is a general statement of purpose; objectives are specific, measurable actions or intentions that lead to achieving the goal(s).

Examples:

- Goal statement: to identify and assess the nature and extent or threat of NPS pollution on reservation lands and waters

- Objective: to quantify and qualify the impairments to tributary X from pollutant Y

The introduction is an excellent place to include maps of reservation lands and waterways. It is vital to establish a spatial context for pollution control activities, and maps are one of the quickest ways to do that. A map will allow the reader to quickly see where impaired waters are and what areas and communities they will affect. Tribes can include a regional map, a map of reservation lands, or a map of the watershed. If the tribe does not have map-making or geographic information system (GIS) capabilities, look for partners that can help, such as local universities.

Example from the Santa Ynez Band of Chumash Indians (SYCR 2006a)

The Santa Ynez Chumash Reservation is located in the south-central portion of Santa Barbara County, California. The Tribal lands consist of 148.28 acres in the Santa Ynez Valley and are inhabited by approximately 287 residents. The Santa Ynez Band of Chumash Indians (SYBCI, or Tribe) is annexing an additional 6.9 acres from Santa Barbara County for development of a museum, cultural center, and park. Zanja de Cota Creek, a perennial tributary to the Santa Ynez River, runs through the length of the Reservation. Residences flank the Creek on both sides, and a small amount of commercial development exists in the northern area. The Reservation sits atop the Upper Santa Ynez groundwater basin, which supplies a portion of drinking water to the Santa Ynez River Water Conservation District. The Tribe is concerned with improving and maintaining surface water and groundwater quality for future generations, as well as for the Zanja de Cota Creek watershed.

Accordingly, the Santa Ynez Band of Chumash Indians began the process of developing a Water Pollution Control Program (WPCP, or Program) in accordance with the Clean Water Act (CWA) and the Tribe's original goals of ensuring fishable, swimmable, and safe waters. The ultimate goal of the WPCP is the development and implementation of water quality standards for future protection and sustained use of valuable Reservation water resources, protection of public health and welfare, and the enhancement of water quality.

Example from the Suquamish Tribe (ST 2008)

The goal of this assessment and management program is to create a general reference which the Suquamish Tribe can use to coordinate and maximize the effectiveness of its internal and external efforts to prevent, reduce and mitigate nonpoint source pollution of the waters within and adjacent to the Port Madison Indian Reservation. The objectives of the assessment and management program are: (1) Provide a description of the present status of Reservation waters, (2) to describe some of the processes that have a deleterious impact on those waters, and to (3), outline a range of options that can address current and foreseeable negative impacts. We understand that funds provided through section 319 of the Clean Water Act are to be used only to address nonpoint source pollution as it impairs or threatens the quality of Reservation waters.

The Port Madison Indian Reservation was established for the Suquamish Tribe under the terms of the Treaty of Point Elliott of 1855. The 7,392 acre Reservation is divided into "fee land" and "trust land." Fee land, also known as "fee simple," is land within the external boundaries of the Reservation that was principally sold to non-Indians and taken out of trust status under the General Allotment Act; fee land can be owned by Tribal members or other Native Americans. Trust land is held in trust by the federal government for the benefit of the tribe or tribal member; this land cannot be subject to municipal, county, state, or federal taxation.

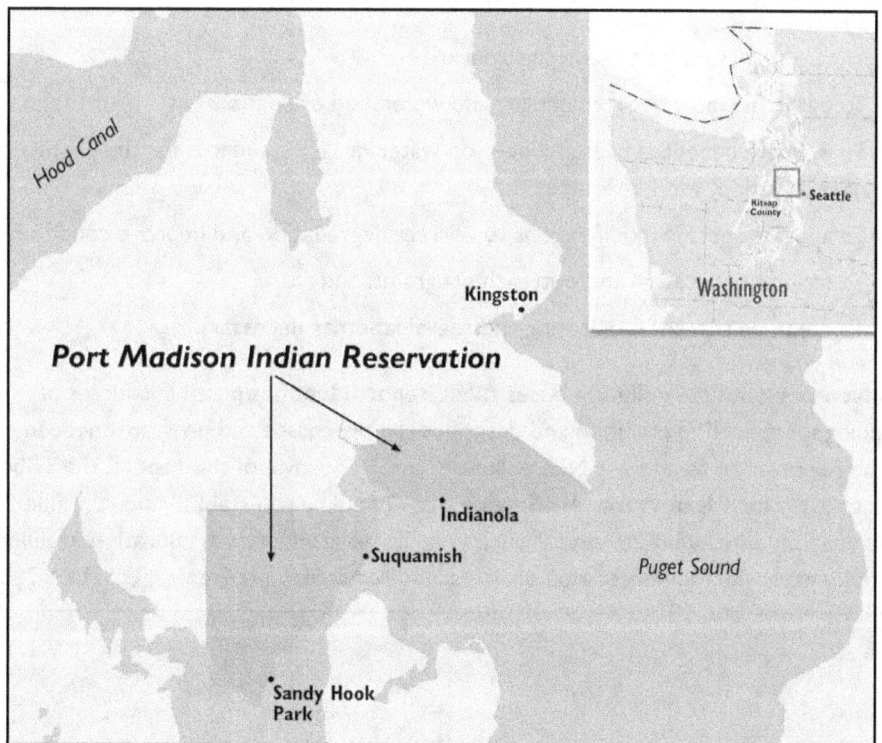

Figure I-2.   Location of Port Madison Indian Reservation.

The general water use and criteria classification of surface waters within the basin has until recently been classified as AA (extraordinary). Characteristic uses for AA waters include water supply for domestic, industrial, and/or agricultural purposes; salmon and other fish rearing, spawning, and harvesting; shellfish (i.e., clam, oyster, mussel, crab) rearing, spawning, and harvesting; wildlife habitat; and, stock watering. The revised water quality standards for Washington State now simply regard the waters of the PMWRB as protected for the designated uses of: salmonid spawning, rearing, and migration; primary contact recreation; domestic, industrial, and agricultural water supply; stock watering; wildlife habitat; harvesting; commerce and navigation; boating; and aesthetic values (WAC 173-201A-600). This change appears to relax the dissolved oxygen and temperature criteria DO was 8.5 mg/L and is now 8.0 mg/L; temperature was 16 °C and is now 17.5 °C.

The climate of the study area is a marine zone with narrow temperature ranges and moist air. The average daytime low and high temperatures of the area ranges from 30s to 40s °F in the winter and 70s to 80s °F in the summer. The average annual rainfall is approximately 35 inches, with about 70% of the precipitation occurring during the period of October to March. Streamflow characteristics in the basin closely correspond with seasonal precipitation patterns. High flows are usually found during the higher precipitation periods of November to March. The lowest streamflows are normally found during the period of July to September (USGS 1979).

The primary objectives to achieve this goal are

- To perform water quality sampling and watershed evaluation;
- To evaluate beneficial uses and develop water quality standards for the protection of reservation water resources;
- To address specific modifications to correct degradation and improve conditions;
- To facilitate education and outreach programs; and
- To continue current monitoring and reevaluation as necessary.

This Non-point Source Pollution Assessment Report identifies possible sources of nonpoint source (NPS) pollution and describes the processes and programs needed for the Reservation to address NPS pollution. Upon approval of this report, the Tribe plans to apply for Clean Water Act Section 319(h) funding to establish and continue these programs, including "technical assistance, financial assistance, education, training, technology transfer, demonstration projects, and regulatory programs" (U.S. EPA 2000).

Methodology

The methodology section describes the assessment methods used. The information in this section should include:

- How and when field data were collected (surface and ground water)
- Timelines (years in which data were collected)
- Spatial analysis units if GIS was used
- Sources of historical data (for example, from state or other federal agencies)
- The level of the quality of your data
- Sampling design
- Sampling parameters
- Standards used
- Any observations or assumptions made during data collection or analysis
- Data management

Example from the Confederated Tribes of Grand Ronde (CTGR 2008a)

The assessment area encompasses the Willamina Creek and Upper South Yamhill River 5th-field watersheds with an emphasis on the Grand Ronde Reservation and Tribal Trust lands within these watersheds. Data used for the analysis of nonpoint sources of pollution within the assessment area came from a variety of sources, depending on the parameter. Since the Tribe has not established water quality standards of its own, the State of Oregon's water quality standards are used. Included are:

- **Streamflow.** U.S. Geologic Survey streamflow gauging of Willamina Creek (#14193000) and South Yamhill River (#14192500). Both of these stations are below Tribal Trust lands. Data displayed are monthly averages of July, August, and September flows from 1934 to the early 1990s. No other long-term flow monitoring stations exist within the assessment area.

- **Water temperature.** Continuous summer monitoring by the Tribe at 32 sites and by the Yamhill Basin Council at 3 sites between 1999 and 2003 within and near Tribal lands. In order to obtain representative values for all stations for a common year (2003), station values missing in 2003 were estimated using a correlation between 2003 and 2001 values.

- **Bacteria.** Monthly data for three stations (DEQ) in or near the assessment area from August 1986 to February 1988. Stations include; #10954 South Yamhill River upstream of Grand Ronde (near Midway), #10969 Willamina Creek downstream of

Willamina (near mouth), #10951 South Yamhill River at Rock Creek Road (upstream of Sheridan). Monthly data for five Tribal testing stations in the assessment area from March through June 2008. Stations are located on Coast Creek (as passes out of Parcel #1), Cosper Creek (as passes through Parcel #28), South Yamhill River (as passes through Parcel #23), Agency Creek (as passes through Parcel #12), and Wind River (as passes out of Parcel #1).

- *Turbidity.* Monthly data for three stations (DEQ) in or near the assessment area from August 1986 to February 1988. Stations include; #10954 South Yamhill River upstream of Grand Ronde (near Midway; upstream of Tribal Trust lands), #10969 Willamina Creek downstream of Willamina (near mouth), #10951 South Yamhill River at Rock Creek Road (upstream of Sheridan).

- *Nutrients.* Total phosphorus and nitrate nitrogen at monthly intervals for three stations (DEQ) in or near the assessment area from August 1986 to February 1988. Stations include; #10954 South Yamhill River upstream of Grand Ronde (near Midway), #10969 Willamina Creek downstream of Willamina (near mouth), #10951 South Yamhill River at Rock Creek Road (upstream of Sheridan). Monthly data for five Tribal testing stations in the assessment area from February through June 2008. Stations are located on Coast Creek (as passes out of Parcel #1), Cosper Creek (as passes through Parcel #28), South Yamhill River (as passes through Parcel #23), Agency Creek (as passes through Parcel #12), and Wind River (as passes out of Parcel #1).

- *Macroinvertebrates.* Fall monitoring by the Tribe from 1998 to 2002 at 35 sites; all sites were on or upstream of Tribal Trust lands. Not all sites were sampled each year. Results used in this report include general biotic integrity, EPT taxa richness (E = Ephemeroptera, P = Plecoptera and T = Trichoptera), and percentage of organisms that are intolerant EPT.

Some of this information might be included in your Quality Assurance Project Plan (QAPP), which should also be referenced in this section. Provide a timeline and explain how the data were collected (for example, field-collected or observational), the spatial analysis units, the sampling design, the parameters measured, and the standards/narrative criteria employed. Describe whether the tribe has its own standards or uses standards from another agency. Tribes could also use criteria from other sources based on the literature (e.g., nutrient levels for wild rice production).

Example from the Santa Ynez Band of Chumash Indians (SYCR 2006a)

The Tribe is currently developing water quality standards with which to compare surface water monitoring results; they are also developing beneficial uses, numeric and narrative criteria, and an Antidegradation Policy. Eventually, Tribal narrative and numeric criteria will provide the Tribe with its own objectives for Reservation water resources. For the purpose of this assessment report, Central Coast RWQCB standards (State Water Resources Control Board 1994) have been utilized for evaluating the impact of nonpoint source pollution.

A habitat assessment and water quality sampling were performed in January 2005 by the Tribe. A historical review was conducted, which included (1) discussion with Tribe members and Tribal elders; (2) a review of historical photographs to determine changes in land use and fluctuations in the riparian canopy, and to identify significant mass wasting events within the watershed; and (3) a review of data collected since May 2004 from surface water monitoring at the Chumash Wastewater Treatment Plant for the National Pollutant Discharge Elimination System (NPDES) permit number CA0050008. The physical habitat survey was performed at locations ZDC-2, ZDC-3 and ZDC-5, and summarized in detail in the *Draft Final Preliminary Water Quality Assessment Report, Zanja de Cota Creek* (SYBCI 2005a).

The survey included the following elements:

- ■ Use of geographic information system digital elevation data to delineate watershed boundaries and to map key features within the watershed;

- ■ A walking reconnaissance of the entire length of the Zanja de Cota Creek from its confluence with the Santa Ynez River through Tribal lands to the community park above State Route (SR) 246;

- ■ A driving tour with several stops to evaluate headwater conditions northeast of the City of Santa Ynez, land use, and other potential impacts within the watershed; and

- ■ A review of historical photographs to determine changes in land use, fluctuations in the riparian canopy, and to identify significant mass wasting events within the watershed.

Monthly surface water sampling began December 2005, as outlined in the *Sampling and Analysis Plan and Quality Assurance Project Plan* (SYBCI 2005b). These water quality data are collected to create baseline standards for the Creek and to identify watershed characteristics, potential health concerns, and possible detection of contaminants within Zanja de Cota Creek for an overall assessment of water quality and system health. Results from these events are summarized in the *Draft Final Preliminary Water Quality Assessment* (SYBCI 2005b) and in Sections 4.0, 5.0 and 6.0 of this report. Results of continued water quality monitoring will be discussed in quarterly monitoring reports.

Monitoring includes the collection of *in-situ* parameters (DO, SC, temperature, turbidity, and pH) and grab samples from four sampling locations, ZDC-3, ZDC-5, ZDC-7, and ZDC- 8, which are analyzed for total dissolved solids (TDS), TSS, chloride, sulfate, phosphorus, total Kjeldahl nitrogen, nitrates, nitrites, ammonia, dissolved metals, boron, sodium, and bacterial indicators (total coliform, fecal coliform, and *Enterococcus*). After collection, the grab samples are preserved with ice and transported to the appropriate State-certified laboratory. Standard U.S. EPA analytical methods are used for all grab samples collected (SYBCI 2005).

Surface water samples taken from the January 2005 and current monitoring events were compared to the water quality objectives set by the Central Coast RWQCB within the Basin Plan. The levels set for TDS, chloride, sulfate, boron, sodium, nickel, and fecal coliform are based on yearly averages; therefore, a year's worth of data needs to be collected to effectively compare the results. Constituents such as TSS, phosphorus, and dissolved metals (cadmium, chromium, copper, lead, mercury, and zinc), except nickel, have no maximum level set by the Central Coast RWQCB and will be contrasted between sampling locations and sampling events.

A comprehensive and detailed methodology will give your study credibility. Accurately reporting the methods will validate the data in the eyes of the reviewers. The tribe needs to show that it has collected samples using predetermined methods to ensure the accuracy of results.

Land Use Summary

The land use summary explains existing land uses and the characterization of ecological conditions on reservation lands. When possible, use maps and identify watersheds at the 12 Hydrologic Unit Code (HUC) level, describing any subbasins appropriate to your assessment and management needs. Some of the physical characteristics to include are acreage, predominant land types and uses, topography, and information on soils and general geology. Be sure to highlight any characteristics or trends that affect water quality; for example, Karst topography (which allows runoff to easily enter ground water sources), a tendency for alternating flash flooding and droughts, or porous soils that trap and hold pollutants.

In addition, include a description of land uses and socioeconomic conditions in this section. Place emphasis on characteristics that factor into water quality, such as population density, economic activities, or unique challenges faced by residents. Use online resources, EPA Regional staff, and local or tribal authorities to gather information for this section. Some tribes have successfully partnered with state agencies, universities, volunteer groups, and other entities to gather land use and water resource information. Such partnerships help to lay the groundwork for later efforts to assess and manage water quality in areas where tribal and nontribal lands intersect.

For this section and subsequent sections, EPA recommends that tribes present data in charts or tables to the extent possible. However, do not force the information into a format for which it is not suited. Tables and charts are recommended because they are an easy way to present a lot of information to the reader.

Example from the Ute Mountain Ute Tribe (UMUT 2005a)

In the Four Corners region, rangeland and forest account for roughly 85 percent of the entire area, and they cover large areas of the Ute Mountain Ute Reservation as well. Most of the Ute Mountain Ute land is either non-commercial timber land (forest) or rangeland used for open grazing (Table 1-2). The Weeminuche Construction Authority uses several acres as an equipment yard for storage and maintenance of equipment and construction materials. Other uses include recreational use (e.g., Tribal Park), resource extraction activities, and irrigated agriculture. Outside of Towaoc, urban land use is essentially non-existent.

Accordingly, primary land uses on the Ute Mountain Ute Reservation include housing for tribal members, oil, natural gas, and sand and gravel extraction, grazing for Tribal livestock, and the Farm and Ranch Enterprise south of Sleeping Ute Mountain. In addition, the Ute Mountain Utes operate several tourism facilities, including the 125,000-acre Ute Mountain Tribal Park, the Ute Mountain Casino Hotel/Resort, the Sleeping Ute RV park, and Ute Mountain Pottery. Table 1-2 summarizes the current land use on the reservation.

Table 1-2. Land use in Ute Mountain Reservation

Use		Area (acres)
Irrigated farm land:	Farm and Ranch Enterprise	7,127
	Mancos Creek Farm	157
Timber land:	Commercial	0
	Non-commercial	163,767
Livestock Range		401,433
Other uses (non-agricultural)		1,614

Source: Tribal Land Use Commission, as cited in Ute Mountain Ute Tribe, 1999a.

The Ute Mountain Ute Tribe Farm and Ranch Enterprise is an irrigated agricultural project designed for 7,634 acres of Ute Mountain Reservation land in southwest Colorado (UMU 1999b). In addition, the Ute Mountain Ute Resources Department operates the smaller Mancos River Farm, which irrigates a few hundred acres. The Farm and Ranch Enterprise grows triticale and alfalfa hay and small grains including corn, wheat, and barley. The Mancos River farm grows hay and provides irrigated rangeland.

The Farm and Ranch Enterprise primarily grows crops, but also owns ~1,200 head of cattle. The purpose of the project is to operate a profitable agricultural enterprise, in addition to providing skilled year-round employment to Tribal members. The enterprise was established, in part, following a dispute in the 1950s over the completion by the Bureau of Reclamation (BOR) of a project that diverted water away from the reservation to non-Indian ranches. Settlement of the water rights issues raised by this project eventually led to the creation of the Dolores Project and Ute Mountain Ute Farm and Ranch Enterprise.

The Farm and Ranch Enterprise uses water entitled to the Ute Mountain Utes by the Colorado Ute Water Settlement Act of 1988, which facilitated the importation of water for irrigation, municipal and industrial, recreation, and wildlife uses. The Dolores Project is a water storage and delivery project that resulted, in part, from the water rights settlement. Water is stored in McPhee Reservoir, located 10 miles north of Cortez, Colorado and 20 miles from the Ute Mountain Ute Reservation. Water for irrigation, wildlife and recreation is transported from the reservoir through the Towaoc Highline Canal, and municipal water is transported by pipeline from Cortez to Towaoc. The Farm and Ranch Enterprise is designed to encompass roughly 7,600 acres of irrigated cropland, primarily south of Sleeping Ute Mountain, and to use on the order of 23,000 acre-feet per year of water.

Oil and gas leases cover 61,745 acres in the south and east part of the reservation, 54,195 acres of which are actively producing (UMU 1999a). An additional 290,000 acres of reservation is available for oil and gas exploration and development. The lands in Utah consist mainly of residential use and livestock use. Traditional plant gathering and limited gardening is practiced in Allen Canyon, the historical home of the Tribal Members who now live in White Mesa.

Example from the Red Lake Band of Chippewa Indians (RLBCI 2008a)

The following land use summary begins with a general narrative of land use organized by watershed, followed by a pie chart with the most common land uses (>1%) included as groups. Detailed maps with information about land use in all watersheds flowing onto and off of the diminished Reservation are shown below each narrative.

Shotley Brook Watershed

Shotley Brook is mostly an undeveloped watershed, primarily consisting of wetlands with interspersed upland forest stands. Basic land use is summarized in Table 1-3 and Figure 1-3.

Table 1-3. Basic land use patterns in the Shotley Brook watershed

Area: 34,938 acres					
Data Source: Red Lake DNR Land Use (Unpublished)					
Wetland/Open Water	**Forest**	**Farmland**	**Urban**	**Residential**	**Other**
22,237 acres	9,617 acres	1,812 acres	181 acres	40 acres	1,050 acres

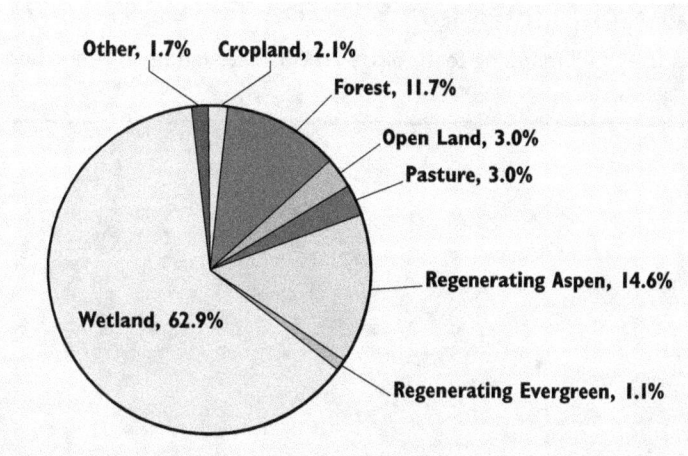

Figure I-3.   General Shotley Brook watershed land use.

An important component of the land use summary is identifying your NPS categories at the subcategory level. Doing this will enable you to focus your remediation efforts and select appropriate BMPs. Table I-4 illustrates how you might summarize that information in this section. Note that the subcategories are self-described by the tribe.

Table I-4. Categories and subcategories of NPS pollution identified in assessment plan from the Red Lake Band of Chippewa Indians (RLBCI 2008a)

Category	Subcategory	Impairment level*
Acid Mine Drainage		4
Agriculture	Crop-related	2
	Grazing-related	2
	Animal holding areas	1
	Streambank erosion	3
Forestry	Vegetated buffers	4
	Streambank erosion	3
Hydromodification/Habitat Alteration	Channelization	3
	Vegetated buffers	3
Marinas/Boating		4
Roads, Highways, and Bridges	Maintenance runoff	3
Urban	Construction	3
	Stormwater runoff	3
Wetland Riparian Management		4
Other	Failing septic systems	4
	Illegal dumping	3

* Scale of Impairments:
Level 1. Confirmed impairment currently exists.
Level 2. Possible impairment: not yet confirmed by monitoring data.
Level 3. NPS pollution occurring with no current impairment to waterbodies.
Level 4. No known NPS pollution occurring or impairment to waterbodies at this time.

Example from Jamestown S'Klallan Tribe (JST 2007)

Figure I-4 uses GIS mapping technology to illustrate land use in the Sequim-Dungeness Valley.

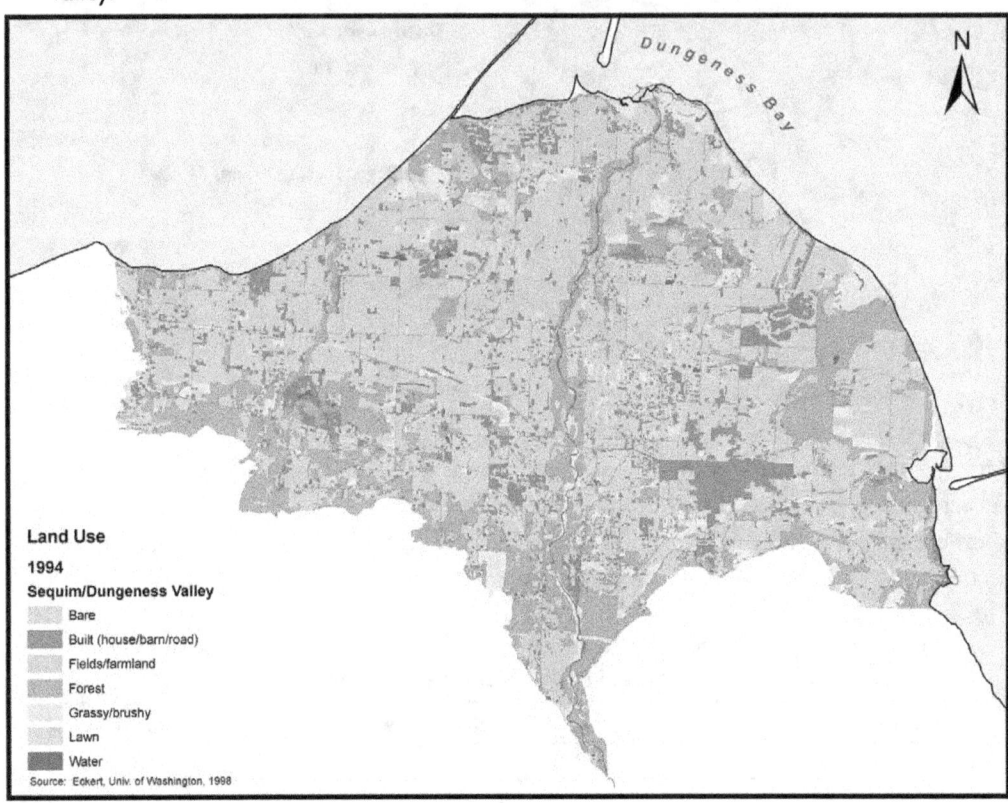

Figure I-4. Land use in Sequim/Dungeness Valley (JST 2007).

Surface and Ground Water Quality Summary

The intent of this section is to present a summary description of the condition of surface water and ground water on the reservation in terms of size, hydrology, and use. All water should be described, including rivers, creeks, lakes, reservoirs, wetlands, and canals. Be consistent in characterizing each waterbody: If it is a stream, call it a stream each time the waterbody is discussed. Frame the discussion of the waterbodies around the hydrology, water quality, and flow conditions. Include intermittent and ephemeral waterbodies as appropriate. Give the length of stream/river miles and lake acreage within the reservation boundaries or areas under tribal jurisdiction. Ground water should also be described.

Detailed maps and other graphics should be used to display the information if possible, including those freely available from Web sites. Including a few pictures of reservation waterbodies in this section is highly recommended. The following example, from the

Jamestown S'Klallam Tribe's watershed-based plan, shows the Dungeness Tribe's watershed planning area boundary (Figure I-5).

Figure I-5. Dungeness watershed planning area (JST 2007).

This section should include enough information to create a picture of reservation waters for the reader. Presenting waterbody information on a watershed basis is encouraged. Ideally, the information is presented in a general way for the entire reservation and in detail for each watershed. It is important to include the following:

- Water types (e.g., river, stream, wetland, reservoir, lake, etc.)
- Waterbody size/length
- Watershed size
- Pollution types (this includes existing pollution and potential pollution)
- Usage information (i.e., cold-water fishery or recreation)
- Historical water quality information, if available

EPA recommends including hydrologic unit code (HUC) numbers in watershed descriptions. A HUC is a 2- to 12-digit number assigned by the USGS as part of its surface waterbody classification system. The United States is divided and subdivided into successively smaller hydrologic units, which are classified into four levels: regions, subregions, accounting units, and cataloging units. The hydrologic units are nested, from the smallest (cataloging units) to the largest (regions). Each hydrologic unit is identified by a unique HUC on the basis of the four levels of classification in the hydrologic unit system, *http://water.usgs.gov/GIS/huc.html*. It is easy to obtain your watershed's HUC on EPA's Surf Your Watershed Web site, *www.epa.gov/surf*. Once you have the HUCs, you will be able to link your information with other data storage programs.

Tribes might not need to collect new raw data. The tribal ambient or drinking water monitoring programs, as well as other federal and state agencies might have already collected data that can be used and incorporated into the assessment report. Most of that information is posted on agency Web sites, and can be easily located and downloaded for tribal use. Cite the source of the data in your report.

Example from the Shakopee Mdewakanton Sioux Community (SMSC 2008)

Shakopee Mdewakanton Sioux Community (SMSC) lands, located entirely within the Lower Minnesota River Watershed, contain two shallow lakes, three shallow, perennial and intermittent streams and numerous temporary to semipermanently flooded wetlands. The two lakes and three streams are partially located on SMSC land; most wetlands occur within property boundaries.

Land Department staff has collected surface water quality data since 1999 as part of CWA Section 106 grant funding. Sample sites have remained constant for the most part; however, a few sites have been eliminated while others have been established.

Parameters collected vary by site; all include nutrients: total phosphorus (TP), ortho phosphorus (OP), nitrite+nitrite (NO2+NO3), total kjeldahl nitrogen (TKN), ammonia (NH4), chlorophyll-a; some include heavy the metals of copper, lead and zinc as well as chloride. Land Department staff has also completed MNRAM function and value assessments on both lakes and many wetland areas.

SMSC waterbodies are degraded compared with pristine waterbodies located in the North Central Hardwood Forest ecoregion. Agriculture and urban impacts to lakes, streams, and wetlands are evident by inferior water quality and loss of vegetative diversity and wildlife habitat.

SMSC drinking water derives from the Prairie du Chien-Jordan and Franconia-Ironton-Galesville aquifers. The Public Works Department completed and passed all tests required by the US EPA which includes daily testing of chlorine, fluoride, iron and manganese and regular testing of bacteria, pesticides and other contaminants. Both wells contain few contaminants, and both fully support drinking water use. Land staff designates SMSC groundwater as good water quality.

Results

The purpose of this section is to clearly define the status of reservation waters using presentation, analysis, and interpretation of available data for each waterbody. At the beginning of this section, it might be helpful to reiterate the most important NPS pollution problems or threats on the reservation. Although all significant pollutants should be addressed, special emphasis can be given to the most important pollutants by presenting more data and providing more in-depth interpretations for them. A general status of the waters should be presented in a narrative overview, which should be supported by a table summarizing the status of all waters of the reservation *by waterbody*. The information presented in your narrative section should include the following:

- Waterbody name

- Area or length

- Each major type of biological, physical, or water quality parameter or pollutant measured or observed (e.g., fecal coliform, pH, Biotic Index, sedimentation, flow)

- Any nonpoint causes or sources of concern (categories and subcategories)

- Water quality parameters indicating impairments or threats and the severity of impact or threat (such as slightly impacted, moderately threatened, severely threatened or impacted)

- An assessment of the overall water quality condition (such as good, fair, poor)

- The type of information used to make the assessment (monitored or evaluated)

This section should contain a summary of the data collected. The actual data should be included in an appendix, or a Web site or document should be referenced. Data analysis should be saved for the Discussion section that follows. When presenting collected data, it might be helpful to use subsections, including those for water quality data, beneficial use designations, and use impairment determinations.

Monitoring data from the tribe's water quality monitoring program, as well as data from other sources (USGS, state, others), should be summarized and presented by watershed in this chapter because watersheds are the basis for NPS management planning. Watershed-scale maps would support this approach. Data should be presented in summary form and interpreted in this chapter, including data collected by the tribe, USGS, state agencies, and other entities. These data should be described in terms of date sampled or year/season assessed, number of years data were collected, and reliability/associated QA. It is important to indicate whether there are any significant data gaps, i.e., gaps that lead to major questions regarding the current status or condition of a waterbody.

Water quality standards or goals (approved or draft), if available, should be included in the data tables to assist in making use impairment determinations. Exceedances of maximum values presented in established standards could be presented as maximums, minimums, and averages for each sampling site. The number of exceedances needs to be expressed in relationship to the total number of samples and the sampling period.

The section should conclude with a determination of which waterbodies are threatened or impaired by NPS pollution, and the general nature of the impairment or threat.

Example from the Red Lake Band of Chippewa Indians (RLBCI 2008a)

Waters on the Red Lake Reservation that have been historically monitored are generally only slightly impacted or in pristine condition. Many Reservation lakes and streams have never been sampled. This is due to the fact that the Reservation is extremely large with a water resources staff that is relatively small. It is likely that with continued monitoring new pollutants and sources will be discovered.

The tribe does not currently have official water quality standards in place. Minnesota state and/or EPA standards were used to determine whether waters were polluted or not. Although all major streams have been included in Table I-5, lakes were only included if data existed in our databases.

Table I-5. Results table for streams on the Red Lake Reservation

Waterbody	Miles	Monitored	Known pollutants	Source	Severity
Mosquito Creek	3 miles	No	None	N/A	N/A
Manomin Creek	10 miles	Once (1990)	None	N/A	N/A
Tamarac River	20 miles	1990-2006	None	N/A	N/A
Shotley Brook	8 miles	Once (1990)	None	N/A	N/A
Sucker Creek	3 miles	No	None	N/A	N/A
Battle River	51 miles	1990-2006	P	Possible agricultural input and natural runoff	Unknown
Blackduck River	320 miles	1990-2006	P	Possible agricultural input and natural runoff	Unknown
Hay Creek	17 miles	No	None	N/A	N/A
Wending Creek	8 miles	No	None	N/A	N/A
Mud River	25 miles	1990-2006	None	N/A	N/A
Shemahgun Creek	2 miles	No	None	N/A	N/A
Pike Creek	51 miles	1990-2006	FCB P	Feedlot Natural and agricultural runoff	Unknown
Little Rock Creek	12 miles	No	None	N/A	N/A
Big Rock Creek	11 miles	No	None	N/A	N/A
Sandy River	48 miles	1992-2006	None	N/A	N/A
Red Lake River	193 miles	1992-2006	None	N/A	N/A
Clearwater River	44 miles	1992-2006	TSS	Agricultural Runoff	Unknown

Example from the Soboba Band of Luiseño Indians (SBLI 2007)

The status of known and potential impairments to surface water bodies on or adjacent to Reservation Trust lands is summarized in Table I-6. The chemical quality of surface water on the Reservation, so far as is known, fully supports the existing uses of groundwater recharge, plant and wildlife habitat, and recreation. The over-riding impairment to support of the existing surface water uses on the Reservation is the reduction of flows by off-Reservation diversions and groundwater pumping. This impairment has been considered by the Tribe primarily in the context of water rights.

Little information is available on water quality on the Reservation, although elevated coliform counts have been observed in Indian Creek and the San Jacinto River. There is no evidence to suggest any deterioration in the chemical quality of surface waters entering the Reservation, but few data are available from which to support this assertion. Because of this lack of information, implementation of a formal program for monitoring of surface water quality and flows on a long-term basis is necessary to properly assess surface quality conditions on the Reservation.

Table I-6. Surface waterbody* summary

Waterbody	Stream (miles)	Impairment	Source category	Severity
San Jacinto River	4.5	Coliform	Unknown	Unknown
		Sediment	Natural and roads (erosion)	Unknown
		Pesticides	Agriculture	Unknown
		Nitrate	Waste water recharge (septic systems)	Unknown
		Petroleum hydrocarbons	Casino parking lot runoff	Unknown
		Reduced flow	Off-Reservation diversion and pumping	Severe
Indian Creek	3.8	Coliform	Possibly agriculture (livestock)	High
		Sediment	Natural and roads, including stream crossings (erosion)	Unknown
		Nitrate	Waste water recharge (septic systems)	Unknown
		Reduced flow	Off-Reservation diversion and pumping	Severe
Poppet Creek	2.2	Sediment	Natural and roads (erosion)	Unknown
		Pesticides	Agriculture	Unknown
		Nitrate	Agriculture (fertilizer) and waste water recharge (septic systems)	Unknown
		Unknown	Landfills (unpermitted)	Unknown
		Reduced flow	Off-Reservation diversion and pumping	Severe

* Type of waterbody: rivers and streams on or adjacent to Reservation Trust lands.

Discussion

The purpose of this section is to interpret your results. This is where you discuss the relationship between the information presented in the Results section and the impacts on water quality (by NPS category) on the reservation. NPS pollution sources should be linked to water quality impairments and threats. It is here, for example, where the report could show that the fertilizer runoff that enters Stream X at area A is contributing to high nutrient levels at points B, C, and D.

Identify the categories of NPS pollution (e.g., agriculture, silviculture, construction) that are causing the majority of the impaired water uses, and rank them by the amount of quantifiable impairment. Highlight the waters that are impaired by each category of pollution, and include a description of the relationship between NPS pollution and specific impaired water quality parameters, as well as any subsequent effects.

Finally, include a summary discussion of the interpretation and analysis of the data. In some cases, the linkages between water quality conditions captured through your data and observed conditions (categories and subcategories of NPS pollution) on the reservation lands or upstream might not be clear. In those cases, additional data may be needed to fully assess pollutant sources, or you might need to begin with a *best professional judgment* until those linkages can be better confirmed. Identifying causes and sources of NPS pollution is an ongoing process that becomes clearer with time. This section should conclude with your best determination of waterbody impairments by NPS pollution.

Example from the Red Lake Band of Chippewa Indians (RLBCI 2008a)

Water quality on the Red Lake Reservation is generally very good. Lake TSI values are remaining stable with only a few exceptions that have shown a slight decrease in the past decade. The Red Lakes continue to fall into the Eutrophic category but remain a healthy walleye fishery. A large number of lakes have not been sampled and will require future monitoring.

Fecal coliform bacteria in Pike Creek are increasing the risk of human health issues on the Reservation. The major known input of FCB comes from a feedlot located just outside the Reservation.

High levels of nutrients and total suspended solids in streams are threatening aquatic life in Reservation waters. The major inputs of these pollutants are likely related to off Reservation agriculture and naturally high nutrient levels in organic soils.

Example from the Suquamish Tribe (ST 2008)

The Washington State DOE proposed 303(d) listings provide a succinct way to identify some of the effects of NPS pollution. Several of the streams are currently listed, among them, Grovers Creek, the largest within the PMWRB.

Table I-7 pulls together the stream, impervious area, identified pollutant concerns, typical NPS sources that contribute to the identified concern, and how much the NPS is likely to contribute to degradation of the stream. Red identifies the parameters that are proposed to be listed as category 5.

We have concerns about low flows in Grovers Creek. Table I-8 shows how the Minimum In-stream Flow increments change over time and repeat in the spring and fall of the year. In the leftmost column are the Minimum In-stream Flow increments; the right hand side shows the time periods subject to the specific flows with time moving in a clockwise direction.

Table I-7. Identified pollutant concerns in reservation waters and likely nonpoint sources

Stream subwatershed	TIA %	Indentified pollutant concerns Parameter (303(d) Category)	Nonpoint sources	Likely contribution
Stoljah	17.8	Dissolved Oxygen (5) Fecal Coliform (5)	4000	High
			7600	High
			4500	Moderate
			7000	High
			7550	Moderate
			8920	High
Kitsap	17	Dissolved Oxygen (5) Fecal Coliform (4B)	4000	High
			7600	High
			4500	Moderate
			7000	Moderate
			7550	Moderate
			7800	Moderate
			8710	High
			7700	High
Cowling	14.4	Dissolved Oxygen (5) Fecal Coliform (5)	4000	High
			7550	Moderate
			7600	High
Indianola	13.4	Dissolved Oxygen (5) Fecal Coliform (4B) pH (2)	4000	High
			7600	High
			4500	High
			7550	Moderate
			7000	High
Kleabel	10.4	Dissolved Oxygen (2) Fecal Coliform (5)	4000	High
			7600	High
			4500	Moderate
			7550	Moderate
			7000	Moderate
Grovers	10.2	Dissolved Oxygen (5) Fecal Coliform (5) Temperature (2) Turbidity (2) Low Flow	4000	High
			7600	High
			4500	High
			7000	Moderate
			7550	High
			7800	High
			8920	High
Sam Snyder	9.1	pH (5) Low Flow	4500	High
			7000	High
			7550	Moderate
			8600	Low
			8920	High
Doe-Kag-Wats	6.9		2000	Moderate
			7550	Moderate

Table I-8. Minimum in-stream flows for Grovers Creek

Minimum in-stream flow (cfs)	Period (colored cells indicate periods when the stream basin is closed to further appropriation.)	
5.5	→ → December 1st to April 14th → ↓	
4.5	↑ November 15th to November 30th	April 15th to April 30th ↓
4	↑ November 1st to November 14th	May 1st to May 14th ↓
3.5	↑ October 15th to October 31st	May 15th to May 31st ↓
3	↑ October 1st to October 14th	June 1st to June 14th ↓
2.5	↑ September 15th to September 30th	June 15th to July 14th ↓
2	↑ ← July 15th to September 14th ← ←	

The following template illustrates a suggested method of displaying a summary of your results and the conclusion of your analyses as narrated in your discussion.

Discussion of NPS Pollution/Sources in Watershed/Subwatershed

- *NPS pollution categories and subcategories of concern*
- *Impairments identified from water quality data analysis*
- *Location of NPS problems: In (Name) and (Name) subwatersheds; in marine subwatersheds only; throughout the reservation*

Example: St. Mary's Creek Watershed

NPS Pollution Category: Agricultural Runoff
Subcategory: Cattle Grazing and Ranching Operations

- *Soil slumping on streamside and other slopes in grazing areas (sediment)*
- *Loss of riparian vegetation from cattle grazing in and out of streams (sediment and temperature problems)*
- *Contaminated runoff and direct deposition of manure and urine to streams (pathogens, ammonia)*

Figure I-6. Summary template.

Selection of Best Management Practices

The purpose of this section is to identify how you will choose BMPs to address the NPS issues identified in your assessment report. Tribes should not discuss the implementation of the BMPs in this section because that information will be required in the management program plan (management plan). In this section, the tribe establishes only that there is an *appropriate system* for choosing which BMPs to implement on reservation lands. This section includes

1. **Core participants.** In addition to listing the agency(ies), organization(s), or task force(s) responsible for BMP selection, briefly describe their mission statements and

membership composition. Also identify the level of participation for each agency, organization, or task force. Here you should also discuss any specific programs (e.g., U.S. Department of Agriculture [USDA] cost-share programs) that have been contacted for BMP selection assistance. Types of participation in BMP selection include:

- Technical assistance
- Education
- Demonstration projects
- Financial assistance

Example from the Ute Mountain Ute Tribe (UMUT 2005a)

Best management practices (BMPs) for the control of nonpoint sources of pollution will be selected based on various factors including information provided by the Non Point Source Task Force and decisions made by the Task Force. The Nonpoint Source Task Force consists of representatives from the following Tribal departments and enterprises and government: Environmental Programs Department, Farm and Ranch Enterprise, Energy Department, Weeminuche Construction Authority, Tribal Park, Planning Department, Natural Resources Department, Tribal Council and Executive Director's Office. Non-tribal agencies represented periodically at the Nonpoint Source Task Force meetings include: Bureau of Indian Affairs, USDA-Natural Resources Conservation Service, U.S. Bureau of Reclamation, and the Indian Health Service.

Implementation of BMPs will be accomplished through a number of nonpoint source pollution programs, funding mechanisms, and educational programs conducted by the Tribe in conjunction with federal and state agencies. Some of the federal government agencies that can contribute to a nonpoint source pollution control program include:

- U.S. Department of Agriculture
- Bureau of Indian Affairs, Department of Interior
- Bureau of Reclamation, Department of Interior
- U.S. Environmental Protection Agency
- Indian Health Service

Additional discussion of funding sources and program requirements of each of these agencies is included in the *Ute Mountain Ute Nonpoint Source Pollution Management Program Plan.*

Example from the Soboba Band of Luiseño Indians (SBLI 2007)

Table I-9. Core participants for BMPs (SBLI 2007)

Participant	Role
Tribal Council, Soboba Band of Luiseño Indians	Lead participant. Sets strategic policies and provides legal authorization for activities.
Environmental Department, Soboba Band of Luiseño Indians	Provides operational lead to surface water monitoring and pollution control activities. Conducts and oversees funding, implementation, and evaluation of monitoring programs and BMPs. Conduct and oversee educational programs for pollution reduction.
Public Works Department, Soboba Band of Luiseño Indians	Provide operational lead for road and firebreak construction, repair, and maintenance.
Soboba Water Utilities, Soboba Band of Luiseñio Indians	Provides monitoring of groundwater quality in Tribal wells.
San Jacinto River Watershed Council	Interagency coordination.
Riverside County Fire Control and Water Conservation District	Policy coordination, program review, and fire control coordination.
Lake Hemet Metropolitan Water District	Policy coordination and program review.
Eastern Metropolitan Water District	Policy coordination and program review.
U.S. EPA	Technical and funding resource. Provides oversight of drinking water quality monitoring.
U.S. National Forest Service	Fire control coordination and technical resource.
U.S. Bureau of Land Management	Fire control coordination and technical resource.

2. **Public participation and governmental coordination.** In this section, highlight the use of public participation and public comment in the process of selecting BMPs and any inter/intragovernmental coordination.

Example from the Suquamish Tribe (ST 2008)

The model for the tribal decision making process regarding choosing BMPs most suitable to address each category and subcategory of nonpoint source pollution identified in our NPS assessment is as follows:

1. Identify all BMPs that are appropriate to each type of NPS pollution through research and consultation.

2. Determine which of the above BMPs are suitable for the PMWRB in terms of scale, environment, and existing infrastructure.

3. Determine likely effectiveness of locally appropriate BMPs in reducing NPS loading through research, modeling, and consultation. Rank them based upon likely performance.

4. Consult with other relevant agencies and jurisdictions to determine which of the BMPs may best be used in coordination with their efforts. Develop formal cooperative agreement(s) when indicated. Identify multiple funding options where possible.

5. Determine which BMPs will have the most favorable results per unit cost.

6. Present options to public meeting of tribal council to allow tribal leadership, tribal members and nontribal public an opportunity to consider options, provide comment, and shape the implementation of the proposal.

7. Implement BMP with adequate resources to perform necessary maintenance and monitor performance.

8. Provide regular updates on BMP status and effectiveness for tribal council and other relevant agencies.

Example from the Shakopee Mdewakanton Sioux Community (SMSC 2008)

The above is, of course, an idealized version and it would most resemble our actual process in larger, more significant programs and projects, those requiring the greatest commitment and coordination of resources. For these, the Tribal Council will provide opportunities for public comment and review and may or may not grant approval as warranted. Where other governments are involved, appropriate government to government protocols will be adhered to and will be an integral part of the process. Smaller proposals, such as low impact, inexpensive, site specific projects, or relatively minor publicity efforts that could be accomplished within our base funding level will typically undergo an internal review including the departments of fisheries, natural resources, and community development.

Section 319 of the CWA requires tribes to assess NPS pollution on SMSC surface waters and determine a management strategy for preventing current and future pollution problems. This report will be sent to federal, state, and local agencies including Scott County SWCD, NRCS, Minnesota DNR, MPCA, and the Cities of Shakopee and Prior Lake. Comments and recommendations on pollution prevention and implementation of BMPs will be incorporated into a final version of this document.

3. **Existing BMPs.** Describe existing BMPs and organize them by category of NPS pollution. A table is a straightforward way of listing the existing BMPs (an example table follows).

Table I-10. Existing agriculture BMPs by NPS category

NPS category	Nonpoint source	NRCS conservation practice standards		Partners	Potential funding
Hydrologic and Habitat Modification	Historic Overgrazing, Erosion and Habitat Destruction and Natural Geologic	322	Channel Vegetation	Tribal EPA/ NRCS	CWA 319
		390	Riparian Herbaceous Cover	Tribal EPA/ NRCS	CWA 319
		395	Stream Habitat Improvement and Management	Tribal EPA/ NRCS/ USFW/ University	CWA 319/ USFW
		584	Stream Channel Stabilization	Tribal EPA/ NRCS	CWA 319

4. **Pollution reduction.** Include a description of the process that will be used to select BMPs aimed at reducing the level of pollution resulting from identified nonpoint sources of pollution.

Example from the Red Lake Band of Chippewa Indians (RLBCI 2008a)

Table I-11 shows the BMPs that RLDNR Waters has identified as appropriate based on the pollutant sources addressed in this assessment as well as sources not yet resulting in impairment. These BMPs were chosen with input from RLDNR Forestry and Wildlife programs. The majority of the BMPs selected are intended to prevent NPS pollution rather than to remove it.

Table I-11. NPS category and associated information (RLBCI 2008a)

NPS category	Source	BMP(s)	Responsible party	Potential funding sources
Agriculture	Nutrients	NRCS-590 Nutrient Management	Private-off Reservation	USDA
	Erosion	NRCS-528 Prescribed Grazing	Private-off Reservation	USDA
	Pathogens	NRCS-528 Prescribed Grazing	Private-off Reservation	USDA
	Pathogens	NRCS-634 Manure Transfer	Private-off Reservation	USDA
Forestry	Erosion, Nutrients	MN Voluntary Site-level Forest Management Guidelines	Tribe/Private Logging Operations On and Off Reservation	Tribe, 319, USDA
Hydromodification/ Habitat Alteration	Erosion	Streambank Stabilization	USACE	USACE, 319
	Erosion	Restoration of River Morphology	USACE	USACE, 319
Roads, Highways, and Bridges	Run-off	NRCS-570 Run-off Management	BIA, Tribe, MNDOT	MNDOT, USDA, 319, BIA
Construction	Run-off, Erosion	NRCS-570 Run-off Management	BIA, Private Contractors, Tribe	BIA, Tribe, 319, Contractors

Example from the Santa Ynez Band of Chumash Indians (SYCR 2006a)

The pollutants of highest threat to Zanja de Cota Creek fall under the "Urban" U.S. EPA NPS category. Corrective action will be taken to identify and/or resolve NPS pollution sources. Short-term goals include attempting to locate the source of bacterial inputs in Reservation waters. Long-term goals include continued public outreach and education, continued water quality monitoring, reduced future construction project impacts, and sustained land stewardship.

A microbial source tracking study should be conducted to ascertain the current sources of bacterial contamination so they may be mitigated. Bacterial Source Tracking is a tool that determines the DNA fingerprints of five colonies of *E. coli* and statistically differentiates it as being from human or animal sources. The primary benefits of using this tool are (1) it increases the ability of the Tribe to locate the source(s) of the contamination with some specificity and (2) blind samples can be used (i.e., a separate DNA fingerprint library does not have to be developed for the Zanja de Cota Creek). An additional source tracking tool is to use Human Fecal Virus Tracking. Detection of human viruses in water samples can serve as an indicator of human contamination. The primary benefit to using this tool is its ability to isolate human-borne viral pathogens and provide a very sensitive indicator for potential human illnesses caused by contact with or consumption of contaminated surface water. Existing coliform data should be used to "isolate" trouble areas and then the DNA tracking techniques should be used to determine whether the source of the observed contamination is human (e.g., failed septic tanks or cesspools, leaking sanitary sewer pipelines, recreational use) or animal (e.g., concentrated animal facilities, wildlife, etc.). With this information, the Tribe could implement BMPs to eliminate or contain the source of contamination.

Existing NPS Control Programs

For each category of NPS pollution (e.g., agriculture, silviculture, urban), identify and describe all available programs for controlling nonpoint sources of pollution regardless of whether they are currently being used on reservation lands. These could include pollution prevention, public outreach, and education programs. Make sure to include programs that are beyond current direct tribal use but that affect the area of concern, such as federal, state, and local government programs, as well as any voluntary or nongovernment programs. This information can come from local sources, online sources, or your Regional EPA coordinator. You can also refer to the partners listed in your existing BMP table or your pollution reduction tables.

Existing NPS pollution-reduction programs for reservation lands should be identified and generally discussed in this section. If there are no available programs, the tribe should identify that as a gap or development need. If possible, the tribe should list the efforts being made and any future plans to fill this gap. These programs might be tribal, local, state, or federal programs that deal with NPS management on tribal lands. This information gives the reader an idea of how the proposed activities fit into the water quality work already in progress.

Example from the Ute Mountain Ute Tribe (UMUT 2005a)

Several programs and projects have been undertaken to address nonpoint source impacts to water quality on the Ute Mountain Ute Reservation, and more are currently underway. One project undertaken in 1999 was to plug an old well that was adding approximately 1 ton of salt per day to the lower San Juan River watershed. The well was plugged using CWA Section 106 Special Studies funding, and the salt load was removed from the system. Another project undertaken in 2002 was to reduce erosion in the greater Towaoc, Colorado area, particularly where a fuels-reduction, forest-thinning project had been undertaken. CWA section 106 Special Studies funds were used to purchase grass seed in order to seed the disturbed area to stabilize soils and prevent sediment movement. Despite drought conditions, the grasses took hold and the project was successful.

Figure I-7. Newly planted native grasses (UMUT 2005).

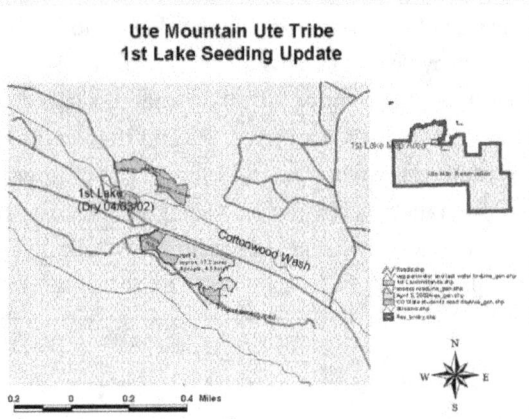

Figure I-8. Map showing First Lake seeding areas (UMUT 2005).

Figure I-7 shows newly planted native grasses (completing with cheat grass) on Special Studies Project— note slash pile to be burned during wetter conditions and steep hill with moderately successful seeding. Previously clear-cut steep hillside was not part of the fuels reduction project, but it was also seeded to prevent erosion. Figure I-8 is the project map for seeding done with the CWA Section 106 Special Studies money to prevent soil erosion after fuels treatment.

Funding from the Bureau of Reclamation is currently being used to enhance two reservoirs that support sport fisheries. First Lake and Horsheshoe Lake have both leaked severely in the past five years, causing the demise of those fisheries. Dirt work has been completed on each reservoir, and a polymer-based sealant or bentonite clay will be used to prevent further leakage in 2004. If this project is successful, fish will be stocked in each lake in 2005 or 2006.

A partnership with the Colorado Division of Wildlife, the Tribe's Brunot Wildlife Department and Environmental Programs Departments, and Mesa Verde National Park has provided a significant ecological restoration to the Mancos River Watershed. The combination of massive, severe-intensity forest fires in the watershed in 2000 and a 5-year drought caused the demise of most of the Mancos River fish. This stream segment is unique because it is populated by almost entirely native fish because of a barricade to migration of San Juan River fish upstream of the Tribe's irrigation diversion dam near

Highway 491/666 in Colorado. An effort was made in 2002 to salvage some of the last Mancos River roundtail chubs—a fish species of "special concern" in Colorado, and listed as threatened in New Mexico. Through a successful captive breeding program, thousands of these fish were returned to the Mancos in September 2003. Also, in April 2004, two other native Mancos River fish species were reintroduced to the river, the flannel mouth sucker and the blue head sucker.

Various other programs have been and are being developed to address nonpoint source pollution. The Tribe's Ground Water Protection Plan was adopted in early 2005. The Ground Water Protection Plan addresses various aquifers, the pollutants and/or land use practices that may degrade the quality of the resources, and how the Tribe intends to prevent that from happening. A major component of the Ground Water Protection Plan is a pesticide management plan that describes preventative measures and how to respond to the detection of those chemicals at various levels. Another preventative measure being undertaken for the protection of ground water in Utah is a sole-source aquifer designation for the White Mesa, UT community drinking water aquifer. This designation will allow the Tribe to undertake more intensive preventative measures to ensure that the aquifer will meet Safe Drinking Water Act Standards. Two monitoring wells, sponsored by the Bureau of Reclamation, have been drilled into an overlying bedrock aquifer in White Mesa in order to monitor any pollution that may emanate from the White Mesa Uranium Mill, 3 miles north of the White Mesa Community of Utes. These wells will intercept any perched ground water that may be affected by the mill, indicate the level of pollution and allow a substantial time to mitigate the situation before any pollution reaches the sole-source aquifer 800–1000 feet below it.

Figure I-9. Tranferring flannel mouth suckers (UMUT 2005).

Figure I-10. Reintroducing fish with tribal/state teamwork (UMUT 2005).

Example from the Red Cliff Band of Lake Superior Chippewa (RCBLSC 2008a)

The Tribe does not currently have a nonpoint source control program in place. The goal is to begin coordination within the Reservation and Tribal Departments, and between the Tribe and other agencies within the Bayfield Peninsula. The Assessment Report and the Management Plan are the first steps in implementing nonpoint source prevention and controls.

Reservation Forestry

The Red Cliff Tribe does not have a Forestry Department. The Bureau of Indian Affairs handles sales of timber. A Forestry Management Plan is currently being drafted by the BIA and Environmental Department. The document will include BMPs to be utilized in forestry on Tribal lands. These BMPs will be incorporated into the Nonpoint Source Management Plan.

Tribal Roads

There are currently no formally adopted nonpoint source control measures. There is occasional cooperation between the Roads Department and the Environmental Department on such issues as weed control measures. Attempts are being made to use native seed mixes for revegetation and the most environmentally safe herbicides for weed control. Waters impacted by nonpoint source pollution are more susceptible to aggressive, non-native species entering the ecosystem. Invasive species can impair quality wildlife habitat, and are very difficult to eradicate once they are incorporated into an aquatic system.

Septic Inventory

A septic inventory is currently being completed and will be assessed by WISCAP. The inventory is being carried out through an EPA grant facilitated by WISCAP with cooperation from Red Cliff staff. The project is entitled "The Red Cliff Septic and Repair Project". After the inventory is complete, suspect systems will be inspected by a licensed inspector to determine which are in need of repair or replacement. The future goal is to obtain funding to implement repair and replacement, thus protecting the Tribe's groundwater and surface water resources.

Conclusions

In this section, provide a summary of the key findings and recommendations of the NPS assessment report by watershed and list special concerns. Identify the categories of NPS pollution that are most detrimental and will be targeted through the section 319 program. Also discuss what is currently being done and plans for the future. This section should provide information on the priority order of the NPS issues on which the tribe will work. Also include a brief discussion regarding how the tribe has met the four eligibility requirements for receiving CWA 319 funding.

Example from the Consolidated Tribes of Grand Ronde (CTGR 2008a)

Formally, the streams in the Upper South Yamhill and Willamina 5th-field watersheds are listed as water-quality limited for flow modification and temperature due to forest practices in the upper watersheds and less protective agricultural practices in the lower watersheds. Despite the listings, evidence from macroinvertebrate and other data suggest that there is still a reasonably healthy ecosystem in place, especially in the headwaters of the two 5th-field watersheds that this assessment covers. When compared to the Portland and Salem metro areas, land use practices are relatively benign and of the type that yet-undiscovered problems are not likely to come to the surface. Management of water quality by the Tribe, using progressive forestry practices and low-impact development in the community, has generally been effective and more innovative than efforts on surrounding private land. So while we are concerned about the listings and do have evidence of reduced water quality lower in the watersheds, there are no major water quality issues within the assessment and none are expected in the near future. However, the Tribe wishes to work with neighboring landowners to modify land use practices that degrade water quality (such as removing or reducing riparian zones) and maintain the areas of existing good water quality.

Example from the Ute Mountain Ute Tribe (UMUT 2005a)

As described in this assessment, various watersheds on the Ute Mountain Ute Reservation have nonpoint source pollution issues. These include erosion and sedimentation, bacteria loading, nutrient enrichment, salinity leaching and loading, selenium enrichment, and radionuclide contamination. Sources and causes of these problems vary from overgrazing and road building to historic mining activities and irrigation of marine shale soils. Each issue will be addressed in the long term using existing regulatory and management programs, a CWA Section 319 Nonpoint Source Management Program, and on-the-ground projects funded by various sources including, but not limited to, CWA Section 319(h) funding. Working with the Nonpoint Source Task Force to identify new problems not identified in this assessment and solutions to them and the issues herein, management changes will be incorporated into day to day operations to minimize and mitigate nonpoint source pollution. Implementation of a CWA Section 319(b) Nonpoint Source Management Program Plan will provide the framework for selection and implementation of best management practices and nonpoint source pollution mitigation strategies.

Nonpoint Source Management Program Plan for 319(h) Eligibility

In this section, you will find
- *Components and format required for a management plan*
- *Recent tribal examples for each component*

The management program plan (management plan) describes how the tribe will use the information contained in the assessment report to address identified water quality impairments and threats. The management plan elaborates on the specific activities to be undertaken to improve or maintain conditions as documented in the assessment report. It is important to note that the CWA requires that the management plan cover at a minimum a 4-year period, with the expectation that it will be updated in the fifth year. EPA encourages tribes to develop management plan periods longer than 5 years, where the first 4 years have specific details and future years are more general and are then updated as part of the fifth-year update. However, if major changes need to be made before the fourth or fifth year, it is acceptable to revise the management plan. Public notice and opportunity to comment are also required for the plan and may be done separately or in conjunction with the assessment report. The section titled Public Notice and Comment, beginning on page I-60, provides more information regarding the topic of public participation.

Components of a Nonpoint Source Management Program Plan

Checklist for Applicants

What you need to have in your management plan

- ☑ Identification of BMPs you plan to use

- ☑ Identification of programs you will use to implement BMPs

- ☑ Schedule for implementing BMPs, with annual milestones

- ☑ Certification of tribal authority

- ☑ Sources of federal and other financial assistance programs

- ☑ Existing programs to ensure no conflict

- ☑ List of local and private experts who will assist the tribe in implementing BMPs

- ☑ Plan should be developed on a watershed basis (encouraged, not required)

This section describes what the CWA requires in a management program plan. The CWA specifically requires the following minimum components in an approvable program plan, and they must be covered in the program plan document. Following this section is a suggested format for writing the plan; the format integrates the required information and allows tribes to describe additional conditions and issues as desired.

1. Identification of BMPs and other measures to reduce NPS pollutant loadings by category and subcategory. The categories and subcategories are those found in your assessment report. Although you should have a list of possible BMPs, you should identify those that will be used to address specific polluted runoff sources.

2. Identification of programs that can help you to implement your NPS management program. These could include, as appropriate, nonregulatory or regulatory programs for enforcement, technical assistance, financial assistance, education, training, technology transfer, and

demonstration projects. At a minimum, the programs listed in this section must also be on the program list in the assessment report, but the list could include other programs to be engaged or developed in the future. This process will assist you in identifying all potential resources available, including those that might not be immediately apparent to the tribe.

3. A schedule containing annual milestones for using the program implementation methods identified in 1 and 2 above. There should be an annual schedule for the first four years. This schedule should reflect the implementation of BMPs identified in 2 above. The schedule should address the most important NPS conditions or gaps in priority order. For example, the most critical sources should be addressed in the first year. In addition to the time frame, each milestone should include the measure/criteria that will be used to document achievement of the milestone.

4. Certification from the tribal legal counsel that the tribe's laws provide adequate tribal authority to implement the program. Any relevant information from the TAS application may be referenced. This certification may also be a part of the management plan submittal or an appendix. This certification is sometimes elaborated on in the text by including descriptions of tribal ordinances, constitutional powers, and the like.

5. Identification of all potential sources of federal and other financial assistance programs and funding that might support your NPS program. You may use the information from the assessment report.

6. Identification of the federal financial assistance programs and federal development projects that affect tribal water resources. These might include individual assistance applications or development projects. Tribes conduct this review to ensure that work being done under the section 319 program is not being undone or duplicated by some other federally funded project.

7. Identification of local and private experts (e.g., range conservationists, fish and wildlife staff, hydrologists, agricultural experts) to be used in developing and implementing a management program. EPA strongly encourages tribes to involve local and public agencies and organizations that have expertise in control of NPS pollution to the maximum extent practicable. This reinforces the importance of water resource partnerships, public notice and comment periods, and the cooperation requirement in the administrative section of the statute (per CWA section 319 (c)).

8. Development on a watershed basis. To the maximum extent practicable, tribes should develop and implement the management program on a watershed basis. For a tribe, this might mean developing a strategy for implementing the program on a watershed basis. EPA's handbook for watershed planning (*www.epa.gov/nps/watershed_handbook*) could be used as a framework to organize activities on the basis of watershed boundaries. Working on a watershed basis would also help tribes when seeking section 319 tribal competitive funding.

Format for the Management Program Plan

Your management program plan should contain the following components, which address each of the numbered plan components described above. Each of the main body components is explained in detail below.

- *Cover Page*—Title and the date (month and year) of the plan

- *Table of Contents (but labeled simply as* Contents*)*—A listing of the major sections of management program plan, lists of figures and tables, appendices, and corresponding page numbers

- *Text (body of the report)*—According to the headings of each major section of the management program plan as listed in the table of contents

  - *Overview*

  - *Introduction*

  - *Summary of Tribal Management Program*

  - *Contents of the Management Program Plan*

    - *Identify possible BMPs, programs, and funding to support your implementation activities. (As part of this section, identify watershed-based activities, if done).*

    - *List of local and private experts who will help you to implement your program*

    - *Schedule for BMP implementation*

    - *Certification of tribal authority*

- *References/Sources of Information*

- *Appendices*

- *Acronyms and Abbreviations List*

Overview Section

Provide a summary of the water resources, uses, and impairments/threats from NPS pollution from your assessment report. Describe the need for an NPS program and the environmental setting. Are there special features, needs, or cultural issues that will affect your program? Use a map to describe the water resources that will be addressed. Describe where the certification of tribal authority is found. For example, the certification could be found in the TAS documentation, the assessment report or in an appendix. The same approach can be taken for the public notice information.

Example from the Stillaguamish Tribe of Indians (STOI 2008)

The Stillaguamish Tribe of Indians (Tribe) is a signatory to the Treaty of Point Elliot, which secures the Tribe's right to harvest fish. The treaty fishing right includes both the right to harvest fish in the Tribe's Usual and Accustomed (U&A) manner and the right to have fish to harvest. Therefore, the Tribe's treaty rights also reserve the right to protect the habitat upon which fish depend within the Tribe's U&A, which is the entire Stillaguamish Watershed. It is therefore important to the Tribe that fish and fish habitat be protected to their fullest extent in order to assure productivity of the resource and the continued livelihood of tribal members. Healthy fish habitat is dependent upon the quality of the surface water resources that fish inhabit. Nonpoint source (NPS) pollution continues to degrade water quality of surface waters on the Tribe's informal "Reservation" and outside of the "Reservation" but within the Tribe's U&A, which in turn leads to impacts on ecosystems including fish habitat and human health.

The Stillaguamish Tribe has been managing NPS pollution through various programs for well over 20 years. However, only recently has the Tribe begun to assess NPS pollution specifically on Tribal land. This NPS assessment report and management plan serve as the beginning point to provide a greater focus on the management of NPS pollution in Tribally managed waters. The NPS Assessment Report provides an understanding of the major sources of NPS pollution affecting Tribal waters, which include development, forest practices, agricultural practices and hydromodification. This plan will describe the various aspects of the Tribe's NPS management initiatives, including a description of actions, priority for their implementation, and estimated budgeting requirements. Elements of these programs were developed to address NPS pollution problems identified from the assessment of Tribal waters, and to provide long-term, comprehensive planning recommendations for watershed protection.

Example from the Confederated Tribes of Grand Ronde (CTGR 2008b)

The purpose of this document is to describe the components of a nonpoint source pollution management program for lands owned and managed by the Confederated Tribes of the Grand Ronde (hereafter, the Tribe). The management program will help assure that nonpoint sources of pollution do not become a problem on Tribal lands. In addition, the program will assist with coordinating with adjacent landowners to minimize their effects on water quality.

Existing data suggest that Reservation streams have relatively high water quality, especially on forestland where a comprehensive Forest Practices Ordinance has been in effect for many years. The cumulative environmental effects of forest activities are addressed in a Biological Assessment for the 10-year Natural Resources Management Plan (Confederated Tribes of Grand Ronde 2002 and 2002). Best Management Practices on land being developed in the town of Grand Ronde and at the Casino site include a progressive plan to collect and re-use runoff water. A Wetland Management Plan describes the treatment of wetlands that are surrounded by developed land or are soon to be developed (Adolfson Associates 2005).

Recommended additions to the nonpoint pollution management program include collecting more complete and up-to-date water quality data to determine if best management practices need to be altered to avoid nonpoint pollution problems. This monitoring will also provide data to inform adjacent forest landowners about stream protection needs. Existing data indicate that stream temperatures increase rapidly for some streams that flow through adjacent private lands.

Introduction

Describe the goals and objectives for the NPS program. These might be water quality restoration for impaired areas and the protection of high-quality waters, as well as controlling NPS pollution. Identify strategies that the tribe is now employing or will employ to ensure a successful program. Describe the geographic extent of where the program will be implemented. Sometimes a tribe might focus on a portion of the reservation or want to work outside the reservation in a cooperative manner to solve a water quality problem in reservation waters. Describe how both reservation and fee lands will be addressed.

Example from Red Lake Band of Chippewa Indians (RLBC 2008b)

The primary goal of the NPS management program is to prevent and control pollution from nonpoint sources and protect or improve water quality on the Reservation. This will be accomplished through emphasis on preventative education in order to protect and preserve tribal waters. These goals will be accomplished by devoting the Section 319 program funding to the following objectives:

1. Implement watershed Best Management Practices (BMPs) and work toward watershed management in an effort to preserve water quality.

2. Environmental Education

   a) Continue to educate the public by organizing the annual water festival.

   b) Work with Riverwatch.

   c) Develop a citizen/student monitoring program on the Reservation.

3. Develop a general habitat protection policy and environmental protection ordinance.

   a) Ordinances will include shoreline development requirements and riparian zone requirements for development, silviculture, and agriculture.

4. Install detention/treatment ponds, rain gardens, sediment basins, etc., to treat major stormwater discharge points not subject to NPDES requirements.

5. Expand baseline monitoring in areas of current or likely future NPS pollution input.

Example from the Suquamish Tribe (ST 2008)

The goal of the management program is to create a general reference which the Suquamish tribe can use to coordinate and maximize the effectiveness of its internal and external efforts to prevent, reduce and mitigate nonpoint source pollution of the waters within and adjacent to the Port Madison Indian Reservation. The objectives of the assessment and management program are: (1) Provide a description of the present status of Reservation waters, (2) to describe some of the processes that have a deleterious impact on those waters, and to (3), outline a range of options that can address current and foreseeable negative impacts.

We will focus on protecting sub-watersheds and catchments that have the lowest percentage of TIA, and work to preserve the integrity of those stream systems. At the same time, we will prioritize remediation, restoration, and mitigation efforts in the sub-watersheds with the highest percentages of TIA. As the PMWRB is a fairly small area and there can be much variation within each stream basin, protection and restoration efforts may simultaneously be applied to a single stream system.

We will rank the streams themselves based on available data regarding physical, chemical and biological parameters. Our methodology will be that described by Scholz and Booth, (2001), or another appropriate to the characteristics of streams in the PMWRB. They describe an efficient way to evaluate the functioning level of stream reaches in order to better focus management efforts.

Management Program Summary

Describe tribal authorities, resource plans, ordinances, and other policies. Provide a description of the administrative location of the program within the tribal government and relationships with other tribal governance structures. Is there an anticipated staffing structure? Will the program be watershed-based? Why or why not? Describe the approach for using local experts, managing cooperative reporting, and conducting consistency reviews of other federal projects the tribe is undertaking.

Example from the Confederated Tribes of Grand Ronde (CTGR 2008b)

The Tribe's Natural Resources Division manages the nonpoint pollution management program. Programs within the Division include timber and roads, silviculture and fire protection, fish and wildlife, environmental resources, and recreation. The staff responsible for nonpoint pollution control maintains ties to managers that are in charge of infrastructure planning and construction for Tribal lands within the town of Grand Ronde and at the Casino. For infrastructure planning and construction, the primary nonpoint pollution management issue is the routing, collection, and re-use of runoff and other wastewater.

The Tribal Council provides ultimate oversight on decisions concerning nonpoint pollution management. Tribal documents that help guide their decisions on nonpoint pollution management issues include:

1. The Environmental Protection Ordinance (Tribal code 651)
2. Forest Practices Ordinance (Tribal code 6.20)
3. Natural Resources Management Plan (2003 to 2012)
4. Wetlands Management Plan (2005)
5. Noxious Weed Inventory and Management Plan (2005)

No Tribal ordinances currently exist that are specific to nonpoint source management issues related to infrastructure planning and construction. The best management practices outlined by the federal Department of Housing and Urban Development for construction currently serve as the minimal standards that apply to Tribal projects.

The legal authority for the proposed nonpoint source management program is based on a 1934 decision regarding the powers of Indian Tribes and a section of the Constitution for the Confederated Tribes of the Grand Ronde. On October 25, 1934, the Solicitor of the Department of the Interior issued an opinion describing the powers of Indian Tribes that is still cited today. A number of powers in that opinion relate directly to the question of Tribal regulatory authority. Within the Constitution for the Confederated Tribes of Grand Ronde is a section (Article I, Section 1) that describes the jurisdiction of the Tribe.

It reads:

"The jurisdiction of the Confederated Tribes of the Grand Ronde Community of Oregon shall extend, to the fullest extent possible under Federal Law, over all lands, waters, property, airspace, minerals and other natural resources, and any interest therein, either now or in the future, owned by the Tribe or individual members held in trust status or located within the boundaries of the tribal reservation which will be established pursuant to the Grand Ronde Restoration Act, notwithstanding the issuance of any existing or future patent or right-of-way."

Excerpt from the Puyallup Tribe of Indians (PTOI 2008)

The legal authority for the administration of the Tribe's proposed nonpoint management program is the 319 Treatment-as-a-State application that is concurrently under review by EPA legal counsel. This application relied on the 1994 WQS TAS application under Section 303 and 518 of the CWA that was approved by EPA in 1994 and the federal TAS simplification rule. The designated tribal department that is responsible for program implementation is the Environmental Department with three full-time water program staff and the Tribal Natural Resources Director, in coordination with Tribal Fisheries. These staff are also responsible for coordination with agencies. Full program implementation will require the hiring of additional staff.

Identification and selection of BMPs that will achieve the Tribe's goals and objectives outlined above were selected by Tribal Environmental staff with approval from the Tribal Natural Resources Director. The Director reports daily to the Tribal Council. Best professional judgment of the staff and coordination and consultation with tribal and agency biologists, tribal members, and planners have and will continue to validate, and if necessary, adjust BMP selection to address major stressors and sources of impairment on the Reservation and throughout the watershed. The Tribal Natural Resources Director, Water Program Manager and technicians have over 50 years of experience working for the Tribe, addressing environmental issues.

BMP selection and program development depends on land ownership, jurisdiction, and funding. The Tribe recognizes the importance of continuing to acquire lands, particularly in the floodplain and along riparian corridors. Controlling land use in the floodplain and adjacent to streams is crucial to preventing and mitigating nonpoint pollution. The Tribe will provide technical support in the development and implementation of BMPs for those lands and waters that we do not have jurisdiction.

Funding plays a crucial role in BMP implementation. The Tribe's Environmental Department is mostly grant funded, thus resources are limited. Tribal BMP priorities don't always align with available funding. Section 106, and GAP funds are currently used to develop environmental programs, monitor and implement water pollution control projects. Continued reliance on outside sources of funding for NPS program implementation is expected.

Management Program Description

This section describes in detail the scope, structure, and functions of the program, including all work anticipated under the NPS pollution control program.

In this section, describe the NPS pollution categories and subcategories to be addressed and relate them back to the specific information on the nature and scope of the issue presented in the assessment report. Tribes may choose to address each category of NPS pollution in a separate subsection in the plan. Each category should have a goal, as well as short- and long-term objectives, covering the five-year cycle of the program plan. Also outline mid-cycle goals/milestones for the next five to ten years that would provide scope and direction for the program update to be undertaken during the fifth year. It might also be important to establish a long-term set of milestones or outcomes looking forward more than 10 years. Some tribes use their general natural resources management documents to define long-range projections for resource improvements.

Elaborate upon specific actions to be taken for each year in the program plan. The actions proposed should relate to the problems identified in the assessment report. These actions will be your schedule of work and the basis for annual work plans submitted for CWA section 319 funding. Include specific information on the type, location, and size of BMPs to be used and the rationale for these decisions. Identify the entity responsible for implementing the BMPs; for example, tribal staff, NRCS personnel, or contractors. If there are cooperators or multiple funding sources for any of the practices, identify and describe them. Describe any plans for long-term operation and maintenance of the management practices. Discuss how success will be measured, such as the number of BMPs installed, their impact on water quality, and so on; these will become your measurable milestones.

Also, describe your polluted runoff issues in a way that clearly identifies nonpoint sources and differentiates them from point sources that might be controlled under stormwater permits. Ensure that you review other programs that might affect the NPS program.

Example from Ute Mountain Ute Tribe (UMUT 2005b)

General Management Plan Information

The overall goal of the management program is to improve water quality on Ute Mountain Ute lands. By establishing water quality standards, the Ute Mountain Ute Tribe has recognized a goal of ensuring that all water sources meet water quality standards for their designated uses. The nonpoint source pollution management program, in conjunction with other Ute Mountain Ute programs, will contribute to this objective. General program milestones can be seen in Table I-12.

Table I-12. General program milestones (UMUT 2005b)

Activity[a]	Frequency/end year
Submit NPS assessment report to EPA	2005
Submit management program to EPA	2005
Submit application for Treatment as a State for CWA Section 319	2005
Propose NPS Management plan to Tribal Council	2005
Appoint NPS committee and convene first meeting	Assembled in 2003/04 Again in 2005
Update management program as needed and review with NPS committee and Tribal Council	Annually
Submit annual status reports to EPA	Annually
Convene NPS committee to review projects and the overall program and set priorities for next fiscal year	Annually
Incorporate priorities into work plan for NPS program and submit to funding agencies (Tribal, EPA, State, USDA, IHS, BIA)	2005 and beyond

[a] Completion of nonpoint source activities will be contingent on program funding

NPS   = Nonpoint source
EPA   = U.S. Environmental Protection Agency, Region 8
USDA = U.S. Department of Agriculture
IHS   = Indian Health Service
BIA   = Bureau of Indian Affairs

The Ute Mountain Ute *Nonpoint Source Assessment* report lists categories of nonpoint source pollution that have been confirmed or are potential sources, as shown on Table I-13. The specific management programs for these categories will focus on prioritizing pollution problems, identifying appropriate BMPs, and implementing BMP demonstration projects. The following considerations will be used in making final management decisions regarding priorities and BMPs:

- Severity of pollution problem and extent of impairment of beneficial uses
- Potential for effectively addressing the pollution problem, given technical and financial constraints (i.e., optimizing economic benefits)
- Public participation and landowner cooperativeness

Table I-13. Categories of NPS and their applicability to the Ute Mountain Tribe (UMUT 2005b)

EPA NPS category	Impairment to Ute Mountain Ute Reservation waters		Possible impairment to off-reservation ranch water
	Confirmed	Possible[a]	
Agriculture			
Crop-related sources: Non-irrigated crop production			
Irrigated crop production		■	
Specialty crop production		■	
Grazing-related sources	■		■
Intensive animal feeding operations			
Silviculture			
Harvesting, restoration, residue management		■	■
Forest management		■	■
Logging road construction/maintenance		■	■
Construction			
Highway/road/bridge construction	■		■
Land development	■		
Urban Runoff/Storm Sewers			
Nonindustrial permitted			
Industrial permitted			
Other urban runoff	■		
Illicit connections/illegal hookups/dry weather flows		■	
Highway/road/bridge runoff	■		

Table I-13. Categories of NPS and their applicability to the Ute Mountain Tribe (UMUT 2005b) *(continued)*

EPA NPS category	Impairment to Ute Mountain Ute Reservation waters		Possible impairment to off-reservation ranch water
	Confirmed	Possible[a]	
Erosion and sedimentation	▪		
Resource Extraction			
Surface mining		▪[b]	
Subsurface mining		▪[b]	
Placer mining			
Dredge mining			
Mill tailings			
Mine tailings			
Petroleum activities	▪		
Acid mine drainage			
Abandoned mining		▪	
Inactive mining		▪	

Source: USEPA 1997, Table I-3.
[a] Source inferred because the facility or activity is present on or near Ute Mountain Reservation
[b] Use under consideration by tribal council

Priorities for implementation are based on the *Nonpoint Source Assessment for the Ute Mountain Reservation of Colorado, New Mexico, and Utah.* Future priorities may be identified and addressed as they come to light and are confirmed by monitoring data. Specifically, the priorities that have been identified in the assessment are:

Table I-14. Reservation waters: levels of specific impairment criteria (UMUT 2005b)

Waterbody	Issue	Level of impairment/ priority
Navajo Wash	Selenium, Salinity, Nutrient enrichment	SEVERE/HIGH
Cottonwood Wash, UT	Radionuclide contamination	SEVERE/HIGH
All < 8000'	Tamarisk/Russian Olive Infestation	SEVERE/HIGH
Mancos River, San Juan River, Navajo Wash	Bacteria levels	Moderate
Mancos River, McElmo Creek, Navajo Wash	Sedimentation/Erosion	Moderate
McElmo Creek	Nutrient Enrichment	Moderate
Mancos River	Metals—Ag, Cu	Moderate
~50% ephemeral streams	Sedimentation/Erosion	Moderate

Excerpt from Puyallup Tribe of Indians (PTOI 2008)

General Management Plan Information

As discussed in the nonpoint assessment, the major nonpoint impacts on the Reservation are the result of activities associated with urbanization—hydromodification (i.e. channelization, dikes, rip rap), habitat alteration (i.e. siltation of stream beds, shoreline development, alteration of riparian areas), and prolifieration of impervious surfaces (i.e. increasing rate and volume of runoff, flow alterations, construction and road building, soil compaction). These activities have resulted in elevated temperatures, siltation of stream beds, low dissolved oxygen, nuisance plant growth, elevated nutrients (high nitrate), and ubiquitous pathogens. Although the chemical parameters of these streams are altered by these activities, the most profound effects are physical and biological. That is, channel cross-sections are enlarged through widening or downcutting, habitat structure is simplified with loss of pools and riffles and little to no riparian cover, low carbon inputs, and aquatic diversity and abundance is poor. Connection of rooftops, gutters, parking lots, roads, and storm sewers to streams has profoundly altered the hydrograph, with most streams exhibiting "flashiness."

Consequently, the hydraulic connectivity of stormwater infrastructure and impervious surface to streams must be addressed. A shift in the focus of stormwater management toward zero connectivity (or disconnection of impervious surfaces) of stormwater infrastructure to streams is necessary if enhancement and recovery activities are going to be successful. Additional activities (i.e. maximizing infiltration, flow control and treatment, better site design and minimization of soil compaction) will serve to more closely mimic natural runoff patterns and result in less nonpoint pollutant impacts to Reservation streams.

As recommended in the guidance, proposed actions are organized by nonpoint source category. Management measures proposed for each source are comprehensive. In general, activities that result in on-the-ground projects and improvements are the priority. Characterization and monitoring to close data gaps, and identify restoration and retrofit opportunities to restore channel form, processes and riparian complexity will be addressed first. This will be done on a sub-watershed basis. Many proposed management measures are ongoing, including enforcing existing agreements with local jurisdictions (i.e. riparian management), coordination and integration with other federal, state and local agencies and programs (TMDLs, salmon recovery planning and implementation, environmental, permit, and plan reviews and consultation).

Table I-15. Plan for management of hydromodification (PTOI 2008)

NPS	Impacts/pollutants	Management measures
Hydromodification	• Disturb stream equilibrium • Disrupt riffle and pool habitats • Create Changes in stream velocities • Eliminate flood functions to control channel-forming processes • Alteration of streambed elevation • Increase erosion and sediment loads (EPA 2007) • Hyporheoic modulation • Elevated thermal regimes • Uncontrolled stormwater loads • Siltation of stream beds	• Restore sinuosity, floodplain connectivity, and modulate stream temperature by re-connecting historic side-channels • Enforce levee management agreement with ACOE and Pierce County • Develop and implement O&M programs to avoid or mitigate physical and chemical impacts, including instream controls (grade control, band stabilization, levee setbacks, non-eroding roadways, streambank, protection, instream sediment load controls, and vegetative cover • Evaluate impacts of surface water quality and in-stream and streamside habitat during dam operation and surface water withdrawal • Work with Ecology to impose flow duration controls/standards to minimize sediment and other pollutants • Preserve the natural hydrologic conditions and protect sensitive hydrologic features, sediment sources, and habitats where possible • Develop watershed–based (sub-watershed basis) hydromodification management plans(s) to address geomorphic and hydrologic impacts of hydromodification on beneficial uses of streams • Where stream bank or shoreline erosion is a nonpoint source problem, stabilized eroding streambanks and shorelines using vegetative methods wherever possible • Protect streambank and shoreline features such as wetlands and riparian zones.

Urban Growth (including on-site sewage systems, construction, and maintenance of roads and highways	• Altered hydrology— Increased volumes and rate of stormwater runoff, low baseflows
	• Increased downcutting, channel widening
	• Reduce channel and floodplain connectivity
		• Promote and implement management measures that maintain natural hydrologic and geomorphic processes, protect riparian corridors, and integrate stormwater control measures into tribal permitting and environmental review processes, emphasize;
		  o Maintaining natural rainfall-runoff ratios
		  o Maximized infiltration
		  o Protection hydrologically sensitive areas, sediment sources, and sensitive

Excerpt from Red Lake Band of Chippewa Indians (RLBCI 2008b)

Excerpt from NPS Category: Agricultural Pollution

Agriculture is the number one source of NPS pollution in Reservation waters, especially in the Blackduck River, Pike Creek, and Battle River Watersheds. Primary pollutants include bacteria and phosphorus. These pollutants come from lands off the Reservation and are carried to local waters. The tribe has been actively cooperating with the Minnesota Pollution Control Agency (MPCA), Natural Resource Conservation Service (NRCS), and local Soil and Water Conservation Districts (SWCDs) to abate these problems. Agriculture on diminished Reservation lands is minimal, consisting of only a few small family farms (estimated at just over 2000 acres total) and does not contribute significantly to NPS pollution. The Red Lake Band owns and operates a medium sized wild rice operation located outside the diminished Reservation. The farm currently follows many BMPs including making use of rotational grazing, riparian zones, and 950 acres of CRP lands. The NPS management program will assist in the development of farm plans for these small farms to prevent degradation. An updated inventory of these lands and their uses with more specific information about types of agriculture will also be a priority.

Short term goals: Reduce the levels of nutrients and bacteria in Reservation waters in the Blackduck River, Pike Creek, and Battle River watersheds.

Objectives:

1. Work with MPCA to enforce feed lot regulations and promote sound management practices such as prescribed grazing and manure transfer on animal feeding operations on lands adjacent to the Reservation.

2. Work with small family farms located on the Reservation to develop riparian zones on farms adjacent to lakes and streams. Assist in voluntary enrollment of these farms into NRCS programs dealing with nutrient management and prescribed grazing where necessary.

3. Work with NRCS to encourage the use of BMPs on lands located in these watersheds with grazing, animal holding, or crop related activities. BMPs will include prescribed grazing, nutrient management and manure transfer.

Short term goals: Determine the contribution of the Red Lake wild rice farm to the impairments in the Clearwater River and reduce the contribution.

Objective: Work with NRCS, the Farm Service Agency, MPCA, and EPA to determine appropriate BMPs (tiling, sedimentation ponds, etc.) and acquire financial assistance to implement them.

Short term goals: Develop accurate, up-to-date, GIS dataset of agricultural land use on the Reservation.

Objective: Delineate exact boundaries of agricultural land use on the Reservation and determine precise land uses: Bison/Cattle pasturing, specific crops, rotations, etc.

Long term goals: Measurable reductions in bacteria, temperature, and fine sediments (TSS) resulting from agricultural activities.

Funding sources will include federal programs, such as the Environmental Quality Incentives Program, Wetland Reserve Program, and Conservation Reserve Enhancement Program which can be used to promote the establishment and protection of riparian zones on farmland.

Table I-16. Schedule of milestones (RLBCI 2008b)

Activity	Year 1	Year 2	Year 3	Year 4
Develop agriculture GIS for lands on Reservation	X	X		
Develop an education and outreach program to work with farmers on and off the Reservation	X	X		
Assist in application for assistance and implementation of BMPs for agriculture on the Reservation		X	X	X
Monitor Project Areas for NPS Improvements			X	X
Increase monitoring for NPS pollution from off-Reservation sources	X	X	X	X

Excerpt from Suquamish Tribe (ST 2008)

Excerpt from Plan for Urban Runoff/Stormwater

Table I-17. Urban runoff/stormwater (ST 2008)

Code	Description	Primary pollutants	% land use	% NPS pollution
4000	Urban Runoff/Stormwater	All		
	- Municipal	All	8.4	5.2
	- Commercial	All	0.3	2.1
	- Residential (e.g., non-commercial automotive, pet waste, etc.)	All	75	52.5
4500	Highway/Road/Bridge Runoff	All	4	15.4
4600	Post-Development Erosion and Sedimentation	All		Sediments: 95

milestones	Year 1	Year 2	Year 3	Year 4
	Draft CAO code language. Design CAO public process. Assess opportunities for joint public information campaign with a neighboring jurisdiction.	Implement one demonstration project. Conduct at least one joint public information campaign. CAO public process	Adopt CAO.	Implement formal process to review permits and applications with neighboring jurisdictions.

This is by far the greatest contributor to NPS pollution within the Port Madison Water Resources Basin (PMWRB) with more than 50% of the loading of nutrients, sediments, and bacteria deriving from this category. The post-development erosion and sedimentation subcategory may provide as much as 95% of the total sedimentation load of the PMWRB. The following general BMP categories will be broadly used to address aspects of the NPS categories and subcategories listed subsequently. We will describe them in greater detail in this section than later sections unless there are specific applications not mentioned here.

Adequate Regulation and Enforcement

Supplement Tribal Code—Formally establish "Critical Areas" within the PMIR. Formalize tribal project review process. Establish protocols with neighboring jurisdictions to review land use and building permits for entire PMWRB. Improve design specifications for tribally sponsored development projects to address stormwater and other NPS issues. Work with neighboring jurisdictions to implement new standards on PMIR. [The Suquamish Tribe's Legal Department would take the lead in formalizing a project review process, developing language codifying critical areas, and establishing protocols with neighboring jurisdictions. The Department of Community Development (DCD), Fisheries Department, and DNR would serve as cooperating agencies, reviewing projects, permits, and improving designs. 1-1.5 FTE.]

Education

Review efforts already being implemented by local governments—target supplemental materials for residents of PMWRB. These will include current materials for pet waste management, septic maintenance, lawn and yard care, household toxics, and water use. Do the same with programs for business owners and managers, targeting for the PMWRB area. Develop materials to give presentations in order to provide information for policy makers within the tribe and other governments. Develop demonstration projects to simultaneously increase awareness, provide useful information, and provide services to local residents and businesses. These may include providing bins for backyard composting and supplying rain-barrels to capture and slowly infiltrate rooftop rainwater. Additional demonstration projects could include coordinating plant salvage operations and establishing one or more native plant nurseries. [DNR would serve as the lead agency; cooperating agencies would include DCD, Fisheries, Education, and many other departments within tribal government. Funding schedule difficult to estimate as it depends upon success and scale of a wide variety of projects. Possibly start with 1.5 FTE then add 0.5 for each year afterwards ending with 3 FTE in year 4.]

Structural BMPs

Infiltration—Reduce effective impervious area through various BMPs such as using pervious alternative paving materials, establishing infiltration space between impervious surfaces when possible, planting native vegetation, amending and aerating soils to reduce compaction.

Detention—Attempt to simulate natural flood hydrograph, increasing the interval between the onset of precipitation and the flood crest while reducing the height of the crest, Where possible, restore or construct wetlands and multiple pond systems to capture and detain runoff.

Filtration—Where suitable, engineer solutions to filter sediments through processes such as grassy swales, filtration basins, sand filters, and even vaults with commercial stormwater filtration devices.

Retrofit Existing Structures—Divert roof runoff to dry wells and rain-barrels, replace culverts and redesign ditches with smaller scale detention structures, route waters to regraded and revegetated areas designed to temporarily store water.

Riparian Restoration—Reestablish native vegetation in riparian corridors, expand the width of the corridor where possible, fence areas to be protected, and use plantings where possible for streambank stabilization. [DCD would be the lead agency concerning structural BMPs, with DNR, Fisheries, and Legal serving as the primary cooperating agencies. This would probably start off more slowly, at about 0.5 FTE, with 0.5 FTE the second or third year for 1 FTE by the 4th year.]

Public Notice and Comment

As part of the process of submitting the NPS assessment report and management plan to EPA for approval, a tribe is also required to submit documentation that adequate public notice and opportunity for comment were given (See section 319(a)(1) and (b)(1)). Section 319 does not provide specific guidance on providing public notice and opportunity for comment; however, the following recommendations are offered to assist tribes in that process.

In this section, you will find

- *EPA's recommendations on how tribes should conduct public outreach and provide public notice so that others can comment on proposed tribal activities*

- *The requirements for tribes regarding public notice and opportunity for comment*

Tribes should work with Regional EPA staff to decide what is appropriate public notice and opportunity for comment.

EPA Regional staff, through consultation with the EPA Office of Regional Counsel, will work with tribes to determine what constitutes adequate public notice and opportunity for comment on the basis of a particular tribal situation. As a general matter, public notice and comment should be extended to nontribal members living on fee lands within a reservation. In some cases, it would be advisable to extend public notice and comment to entities outside the reservation boundaries if an NPS management plan proposes to implement practices that would very likely affect those outside entities. Another consideration to discuss in determining whether to pursue outside public notification is the benefit the tribe could receive from the process. For example, it could be to the tribe's advantage to provide notice to adjacent properties as a way of promoting its management program and, perhaps, leveraging funding and assistance from other groups doing similar NPS work.

There are no specific rules for determining the length of a comment period. A 30-day comment period might be sufficient, but a longer period might be appropriate as well, depending on the tribal communication structure, the size of the reservation, and whether outside notification will be given.

Tribes must include a statement that they have provided notice and an opportunity for public comment regarding their assessment report and management program and a brief description of how that notice was provided.

Tribes should consult with Regional EPA staff to determine the appropriate time frame for conducting public notice and providing opportunity for comment. For example, conducting public notice after Regional review of the NPS management program could be appropriate. If public comments are received, the tribe must provide the response to comments as part of this documentation.

The recommendations above also apply to cases where tribes have revised an NPS assessment report or management plan. The example below was provided by Red Lake Band of Chippewa Indians (RLBCI 2008).

> A public comment period for the Red Lake Nonpoint Source Assessment and Red Lake Nonpoint Source Management Plan took place between the dates of 3/24/08 and 4/23/08. Public notices were posted at all local businesses and government offices. A copy of the notice and a letter were sent to surrounding local government units and interested parties. Interested parties included local Soil and Water Conservation Districts, Minnesota Pollution Control Agency staff, and Minnesota Department of Natural Resources staff. Although a high level of interest was shown in viewing the documents, only one comment was submitted. The Beltrami Soil and Water Conservation District submitted a letter of support as a comment to the documents.

EPA Regional Approval Process

Each EPA Regional office has a tribal contact who can best describe the process that the Region uses to review assessment reports and management plans, as well as TAS eligibility requirements. For a list of EPA tribal NPS coordinators, see *www.epa.gov/nps/tribal*.

Early in the calendar year, tribes should let the Region know they are interested in becoming eligible for the section 319 program. Furthermore, the intent to apply for CWA 319 funding should already be in the tribe's CWA 106 or GAP work plans if those programs are to be used to work toward 319 eligibility.

Regions can provide comments on draft documents to ensure that the approval on final program documents moves smoothly. This should happen in spring and summer. Remember to build in time for a public notice and comment period, as well as potential need for document revisions to address comments. Regional approvals must be completed by the second Friday of every October if you wish to be eligible to receive funds for work in the subsequent fiscal year. The earlier in the year you have your documents ready, the more likely the process will be completed by the deadlines established by the *Federal Register* notice.

Section 319 Funding Process

Every year, Congress appropriates funds dedicated to the section 319 program for tribes, states, and territories. A portion of the tribal funds is dedicated to funding work plans that have been approved under the base NPS program. Such funds are commonly referred to as *base funds*. The remaining dollars are competed nationally to fund NPS on-the-ground projects. That money is commonly referred to as *competitive funds*. Figure I-11 shows total tribal set-aside dollars for fiscal years 2000 through 2010. The increases in funding levels over the years have reflected the increase in the number of tribes that have entered the program, the demonstrated need, and continued congressional support for the section 319 program in general and the tribal 319 program in particular.

> **In this section, you will find**
> - *The difference between base and competitive funds, and potential dollar amounts associated with each type of funding*
> - *Where to find information on how to apply for funding*
> - *The cost-share/match requirements associated with funding*
> - *How tribes may combine section 319 funding with Performance Partnership Grants (PPGs)*

Base Funding

In 2009 tribes were eligible to receive either $30,000 or $50,000 according to reservation area. Although there are other factors that influence nonpoint sources of pollution, EPA uses land area as the deciding factor for allocation of base funds because NPS pollution is strongly related to land use.

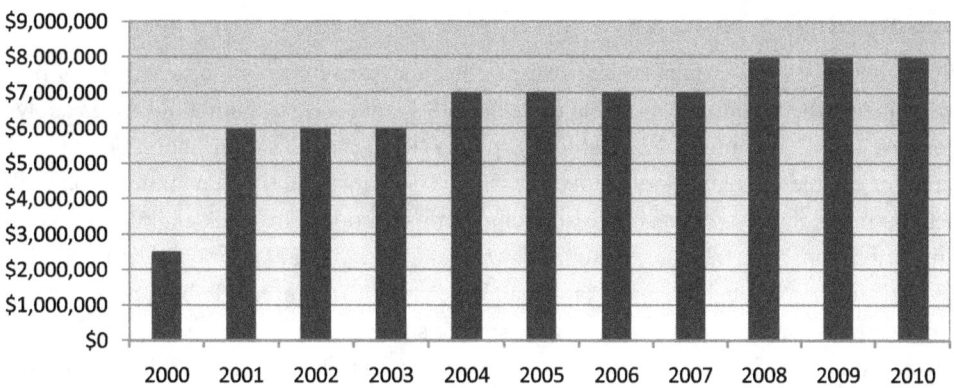

Figure I-11. Total annual tribal section 319 funding.

Base funds are commonly used for funding an NPS coordinator, updating the NPS management program plan, developing a watershed-based plan, or funding activities related to training and outreach. Tribes may also implement on-the-ground projects with these funds if adequate funding exists.

Tribes must refer to the most recent 319 base guidelines when applying for funding because the guidelines are not published annually. They are updated only as necessary. The base guidelines are under *Funding* at *www.epa.gov/nps/tribal*. It is important for tribes to follow application directions to receive base funding.

Tribes are generally required to provide tribal or other nonfederal funds as a cost-share/match for section 319 funds. For purposes here, the terms *cost-share* and *match* are synonymous and do not equate to a dollar-for-dollar match. For information on calculating cost-share/match and determining sources of match, see the section titled Cost-Share/Match Requirements on page I-65, and the Work Plan Development section that begins on page I-68. EPA Regional tribal coordinators or assigned grants project officers are responsible for review and approval of work plans, and they can assist applicants in work plan development. For EPA contact information, see the list of contacts under Part III, Additional Resources for Tribes, or visit *www.epa.gov/nps/tribal*.

Table I-18. Base funding parameters

Area	Maximum funding
Less than 1,000 mi² (less than 640,000 acres)	$30,000
Greater than 1,000 mi² (greater than 640,000 acres)	$50,000

EPA strongly encourages tribes to use either base or competitive section 319 grants to develop and/or implement watershed-based plans containing the nine minimum planning elements (nine elements).

Competitive Funding

The amount of funds available each year for the competitive portion of the tribal 319 program depends on the total funds available for tribal 319 grants and the number of tribes applying for base funds. The annual Request for Proposals (RFP) provides information on the total amount of funds available for competitive tribal 319 grants and the maximum amount of federal funds that an eligible tribe can include in its proposal. In 2009, eligible tribes and intertribal consortia were able to submit a proposal for up to a maximum of $150,000 in federal funding through the national tribal 319 competitive process. Tribes are required to provide tribal or other nonfederal funds as a cost-share/match for section 319 funds. For information regarding cost-share/match, see the section titled Cost-Share/Match Requirements on page I-65 and the Work Plan Development section that begins on page I-68.

Typically, EPA solicits proposals from eligible tribes and intertribal consortia to develop or implement watershed based plans and other on the ground projects, or to develop NPS ordinances that will result in significant steps toward solving NPS impairments or protecting waters from NPS pollution. Detailed information regarding the RFP is not discussed here, because submission requirements and selection criteria can change from year to year. Applicants for competitive tribal 319 grants must refer to the annual RFPs posted to grants.gov at *www.grants.gov* and to the tribal NPS Web site at *www.epa.gov/nps/tribal*. Applicants should carefully read the annual RFP to ensure that their proposals fully address all requirements, including eligibility threshold criteria and evaluation factors. EPA Regions will review each proposal to ensure that threshold eligibility criteria are met. Applicants deemed ineligible for funding consideration as a result of the threshold eligibility review will be notified by the EPA Regions. Proposals that meet the threshold eligibility criteria are submitted to EPA Headquarters for review and ranking by a selection committee. Each proposal is scored by the committee members on the basis of the factors and point system described in the annual RFP. The ranked list is provided to EPA's selection official, who makes final funding decisions.

All aspects of the competitive process are guided by EPA Order 5700.5A1, Policy for Competition of Assistance Agreements. It is important to note that EPA will not discuss draft proposals with applicants, provide informal comments on draft proposals, or provide advice to applicants on how to respond to selection criteria. As part of the annual RFP process, applicants are provided the opportunity to submit questions in writing to the EPA Regional contacts regarding threshold eligibility criteria, administrative issues related to the submission of the proposal, and requests for clarification about the RFP. EPA's responses to questions are posted to EPA's Tribal Nonpoint Source Web site. The Tribal Nonpoint Source Web site is a helpful resource for tribes to access during their proposal preparation.

Cost-Share/Match Requirements

Section 319(h)(3) of the CWA requires that the cost-share/match for NPS grants be at least 40 percent of the total project cost. Match can include:

- Allowable costs incurred by the grantee, subgrantee, or a cost-type contractor, including those costs borne by nonfederal grants

- Cash donations from nonfederal third parties

- Value of third-party in-kind contributions

- Tribal in-kind contributions, such as salary and equipment

To calculate the cost-share/match funds for the total project, the following tables demonstrate a 40 percent (section 319-required cost-share/match), 10 percent (if undue hardship requested), or 0–10 percent (if the work plan is combined in a Performance Partnership Grant [PPG]) cost-share/match on a section 319 base funding request of $50,000 and $30,000.

Table 1-19. Match calculation table for tribes eligible for $50,000 of base funding (> 1,000 mi²)

Total project cost	Nonfederal match	Federal share	Nonfederal Match	Federal share
$83,333	40%	60%	$33,333	$50,000
$55,556	10%	90%	$5,556	$50,000
$52,632	5%	95%	$2,632	$50,000

Table 1-20. Match calculation table for tribes eligible for $30,000 of base funding (< 1,000 mi²)

Total project cost	Nonfederal match	Federal share	Nonfederal match	Federal share
$50,000	40%	60%	$20,000	$30,000
$33,333	10%	90%	$3,333	$30,000
$31,579	5%	95%	$1,579	$30,000

Example Calculation:

a. If you know the total project costs:

(1) Multiply the total project costs by the cost-share/ match percentage needed.

(2) The total is your cost-share/match amount.

For example:

If you are requesting $30,000 of base funding, and your total project cost = $50,000, and you need 40 percent cost-share/match, so $50,000 × .40 = $20,000 (cost-share/match).

OR

b. If you know the total federal funds requested ($30,000 for this example):

(1) Divide the total federal funds requested by the maximum federal share allowed.

(2) Subtract the federal funds requested from the amount derived in step 1.

(3) The amount derived from step 2 is the nonfederal match.

For example:

(1) If the federal funds requested = $30,000 and the recipient cost-share/match is 10 percent, the federal share = 90% or 0.90. $30,000 ÷ 0.90 = $33,333 (total project cost).

(2) $33,333 − $30,000 = $3,333

(3) The nonfederal match = $3,333

Table 1-21. Match funding requirements

Grant request	Cost-share/match required
Section 319 match requirement	40%
Undue hardship	10%
Performance Partnership Grant*	0%–10%

* Different funding in a PPG have different match requirements

Applicants should be aware that funds originating from the Bureau of Indian Affairs may be used as cost-share/match for section 319 (pursuant to 25 U.S.C. section 458cc). These funds are treated as nonfederal funds for purposes of meeting cost-share/match requirements.

EPA's regulations also provide that EPA may decrease the cost-share/match requirement to as low as 10 percent (or in certain circumstances as low as 5 percent with respect to tribal 319 funds that have been included in PPGs). Tribes must demonstrate in writing to the Regional administrator that fiscal circumstances within the tribe (or within each tribe that is a member of the intertribal consortium) are constrained to such an extent that fulfilling the cost-share/match requirement would impose undue hardship (see 40 CFR 35.635).

Performance Partnership Grants

Section 319 grants are eligible to be included in PPGs (see 40 CFR 35.530). PPGs enable tribes and intertribal consortia to combine funds from more than one environmental program grant (e.g., section 106, General Assistance Program [GAP]) into a single grant with a single budget. The purpose of PPGs is to strengthen planning and priority setting through better deployment of resources, flexibility in directing resources to where they are most needed, linking activities to overall environmental goals and outcomes, fostering innovation, and streamlining administrative requirements.

If a tribe includes a section 319 grant as a part of an approved PPG, the cost-share/match requirement may be reduced to 5 percent of the allowable cost of the work plan budget for the first 2 years in which the tribe receives a PPG. After 2 years, the cost-share/match may be increased to up to 10 percent of the work plan budget. This determination of a tribe's ability to meet match is made by the EPA Regional administrator after various socioeconomic factors of the tribe are taken into consideration (see 40 CFR 35.536). The Regional administrator also has the regulatory authority to waive all cost-share requirements for the PPG if socioeconomic factors show that meeting match would impose an undue hardship (see 40 CFR 35.536(d)). EPA is developing a consistent, standardized approach to assist Regional administrators in determining tribal match waiver eligibility. The socioeconomic criteria will be based on publicly available, nationally consistent data.

Recipients are not required to account for PPG funds in accordance with the funds' original funding program, although total PPG expenditures need to be accounted for. If a tribe proposes a PPG work plan that differs significantly from any of the approved 319 work plan components, the EPA Regional administrator must consult with the NPS national program manager before the 319 grant may be included in the PPG. Tribes interested in combining 319 dollars into a PPG should consult with their EPA Regional tribal coordinator to determine whether a PPG is an appropriate option.

Participants at a regional tribal NPS Workshop view native species revegetation project. (New Mexico 2008)

Grant Timeline

The base and competitive timeline has been developed to coordinate proposal submissions and the receipt of fiscal year appropriations, and to avoid busy summer schedules when the bulk of project implementation occurs. Deadlines vary from year to year; however, the timeline in Figure I-12 generally outlines the process.

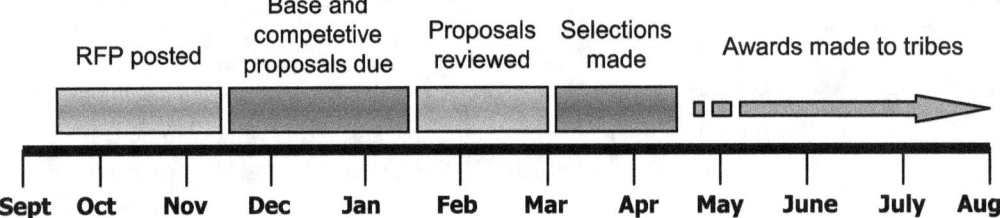

Figure I-12. Grant timeline.

Work Plan Development

Eligibility to receive a section 319 grant is complete once the tribe has obtained TAS and approval of its assessment report and management plan. The next requirement before receiving CWA section 319(h) grant funds is an approved work plan. A work plan outlines the goals and objectives of the grant with a schedule and budget breakdown to be carried out with CWA section 319 grant funds. To begin developing a work plan, refer back to the NPS assessment report and determine the priority NPS problems to be addressed in the upcoming grant cycle. Then refer to the NPS management program plan to determine the BMPs to implement to address the priority NPS problem identified in the NPS assessment report. The work plan should align with the schedule in the NPS management program. Next, develop a work plan and estimated budget using the priority NPS problem and selected BMPs.

In this section, you will find
- What is a work plan?
- What are the five requirements of a work plan, and what must a tribe do to meet them?
- What does a work plan look like?

All work plans funded with federal grant funds must, at a minimum, include the work plan information required at 40 CFR 35.507. For additional work plan requirements for base and competitive grants, see the most recent guidelines or RFPs online at *www.epa.gov/nps/tribal*.

There are five requirements for a work plan at 40 CFR 35.507:

1. Work plan components to be funded under the grant (you may also show some components that are funded by other sources).

2. Work plan commitments for each work plan component, and a time frame for their accomplishment.

3. Estimated work years and estimated funding amounts for each work plan component.

4. Reporting schedule and a description of the performance evaluation process.

5. Roles and responsibilities of the recipient and EPA in carrying out the work plan commitments.

EPA recommends including a short introduction or background section to remind the reviewer where the activities came from (e.g., assessment report on *X* river or management plan commitments for year *Y*). It is advisable to also provide a brief physical description of the reservation land and waters. EPA project officers change over time, so including that type of information helps EPA to coordinate programs with greater understanding.

In-Depth Detail on the Five Requirements

Work Plan Components

The first requirement is to identify the work plan components to be funded under the grant. Identify the major goals to be carried out in the work plan that addresses the NPS problem (as outlined in the management program plan). A typical NPS work plan would have about three to five components depending on activities to be carried out with the available resources.

Work Plan Commitments

The second requirement is to identify the work plan commitments (tasks) for each work plan component, along with a time frame (milestones) for their accomplishment. Identify the tasks to be carried out that will meet the goals of the work plan. Identify how progress toward carrying out the commitments will be measured. Identify the start and end dates for each component. Identify the anticipated environmental outputs and outcomes, and the plan for tracking and measuring progress toward achieving the expected outputs and outcomes.

Apache Crown Dancer I (Allan Houser Haozous), Albuquerque.

Environmental Results (per EPA Order 5700.7)

The underlying principle of EPA Order 5700.7, Environmental Results under EPA Assistance Agreements, is to ensure that assistance agreements (grants, cooperative agreements) administered by EPA, such as section 319 grants, are results-oriented and align with EPA's strategic goals. Providing linkage to strategic plan goals, as well as developing measurable outputs and overall goal outcomes, provides a way of measuring the environmental benefits that could be achieved through a grant award. Given the importance of this principle, it is required that grant awardees state the outputs and outcomes that are the expected results of their work plan activities/components.

An *environmental output* (or deliverable) is an environmental activity, effort, or associated work product related to an environmental goal or objective that will be produced or provided over a period of time or by a specified date. Outputs can be quantitative or qualitative, but they must be measurable during an assistance agreement funding period. Examples of environmental outputs include:

- Development of a nine-element watershed-based plan
- Miles of fence line installed
- Feet of stream bank planted
- Amount of large, woody debris placed
- Number of stream meanders restored
- Percent reduction in road density

Tribal NPS planning meeting at the Region 6 Tribal NPS Workshop, 2008.

An *environmental outcome* is the result, effect, or consequence that will occur from carrying out an environmental program or activity that is related to an environmental or programmatic goal or objective. Outcomes could be environmental, behavioral, health-related or programmatic in nature; must be quantitative; and might not necessarily be achieved within an assistance agreement funding period. Outcomes may be short-term (changes in learning, knowledge, attitude, skills); intermediate (changes in behavior, practice, or decisions); or long-term (changes in condition of the natural resource). Examples of environmental outcomes include:

- An increased number of NPS-impaired waterbodies that have been partially or fully restored to meet water quality standards or other water quality-based goals established by the tribe

- An increased number of waterbodies that have been protected from NPS pollution

- Increased abundance and diversity of fish or macroinvertebrate species

- Increased NPS knowledge on the part of community members

- Increased knowledge on the part of trained staff in the 319 program

Estimated Work Years and Funding Amounts

The third requirement is to identify the estimated work years and estimated funding amounts for each work plan component. This requirement has two parts. First, identify the percentage of time that a full-time equivalent (FTE) is estimated to carry out each work plan component. For example, a 100 percent FTE is 1.00 work-year. If the employee will carry out work plan component #1 part-time or about 25 percent of a full-time workload, the estimated work-years for work plan component #1 would be 0.25 work-year/FTE.

The second part is to identify the estimated funding amount to carry out each work plan component, which includes salary, fringe, travel, equipment, supplies, contractual, and other expenses. The required cost-share/matching funds may also be included in the estimated funding amount for each work plan component.

Reporting Schedule

The fourth component of the work plan is to identify the reporting schedule and to describe the performance evaluation process, required according to 40 CFR 35.515, Evaluation of Performance. This information will meet the regulatory requirements for

- A discussion of accomplishments as measured against the work plan

- A discussion of the cumulative effectiveness of the work performed under all work plan components

- A discussion of existing and potential problem areas

☐ Suggestions for improvement, including, where feasible, schedules for making improvements

To best address this requirement, the required 40 CFR 35.515 information for the performance evaluation process may be copied directly into the work plan. This evaluation provides the basis for the tribal progress reports.

---

§35.515 Evaluation of Performance

(a) *Joint evaluation process.* The applicant and the Regional Administrator will develop a process for jointly evaluating and reporting progress and accomplishments under the work plan (see section 35.507(b)(2)(iv)). A description of the evaluation process and reporting schedule must be included in the work plan. The schedule must require the recipient to report at least annually and must satisfy the requirements for progress reporting under 40 CFR 31.40(b).

(b) *Elements of the evaluation process.* The evaluation process must provide for:

(1) A discussion of accomplishments as measured against work plan commitments;

(2) A discussion of the cumulative effectiveness of the work performed under all work plan components;

(3) A discussion of existing and potential problem areas; and

(4) Suggestions for improvement, including, where feasible, schedules for making improvements.

(c) *Resolution of issues.* If the joint evaluation reveals that the recipient has not made sufficient progress under the work plan, the Regional Administrator and the recipient will negotiate a resolution that addresses the issues. If the issues cannot be resolved through negotiation, the Regional Administrator may take appropriate measures under 40 CFR 31.43. The recipient may request review of the Regional Administrator's decision under the dispute processes in 40 CFR 31.70.

(d) *Evaluation reports.* The Regional Administrator will ensure that the required evaluations are performed according to the negotiated schedule and that copies of evaluation reports are placed in the official files and provided to the recipient.

Roles and Responsibilities

The fifth work plan component is to identify the roles and responsibilities of the recipient and partners in carrying out the work plan commitments. Specify who will carry out which components and who has the lead to carry out each component.

Table I-22 is a work plan template that includes the five requirements of a work plan per 40 CFR 35.507. For base funding, some EPA Regions require a more narrative work plan than the one shown in the table. For more information, consult with your Regional NPS tribal coordinator.

Table I-22. Sample work plan NPS pollution control program (CWA section 319)

Work plan components with commitments and environmental results	Dates		Outputs/ deliverables	Responsible staff and work-years	Estimated cost*	Status
	start	end				
1 COMPONENT: Decrease livestock access to riverbanks of XX River. Commitment 1(a): Purchase fencing supplies. Commitment 1(b): Fence (X) yards of riverbank. OUTCOMES/ENVIRONMENTAL RESULT: Water quality will be improved through a decline in sediments in the river by preventing livestock access to the river.	10/01 10/30 11/01 12/30		Submit *before* and *after* photos with quarterly report.	WQS 0.20 Crew	$15,000	
2 COMPONENT: Revegetate project area with native vegetation. Commitment 2(a): Purchase trees, shrubs, and grasses. Commitment 2(b): Plant trees, shrubs, and grasses. Commitment 2(c): Install watering system for vegetation. OUTCOMES/ENVIRONMENTAL RESULT: The growth of native vegetation at the project site will stabilize the riverbank and reduce the amount of sedimentation entering the river.	2/1 3/1 4/01	2/28 4/30 5/30	Submit *before* and *after* photos with quarterly reports.	WQS 0.20 Crew	$10,000	
3 COMPONENT: Provide alternative water sources for livestock. Commitment 3(a): Purchase solar panels and troughs. Commitment 3(b): Install solar-powered water system. OUTCOMES/ENVIRONMENTAL RESULT: Providing alternative water resources for livestock will prevent them from going into the river, leading to a decline in sedimentation and pathogen introduction.	11/01 1/01	12/30 2/30	Submit photos of new water system with quarterly report.	WQS 0.16 Crew	$8,000	
4 COMPONENT: Educate tribal community on the importance of water quality protection and NPS pollution control program. Commitment 4(a): Develop water quality education/ outreach program. Commitment 4(b): Hold public outreach meetings and conduct site visits to project area. Commitment 4(c): Contribute articles to the Environmental Department's monthly newsletter. OUTCOMES/ENVIRONMENTAL RESULT: Changes will occur in awareness and understanding of the status of water quality and the effects of NPS pollution problems as demonstrated by direct participation in workshops, meetings, and site visits.	10/01 Semi-annually Monthly	12/31	- Submit to EPA an education and outreach program description. - Include in quarterly reports a summary of public meetings and site visits. - Provide copies of newsletters.	WQS 0.10	$3,000	

Table I-22. Sample work plan NPS pollution control program (CWA section 319) *(continued)*

Work plan components with commitments and environmental results	Dates start	end	Outputs/ deliverables	Responsible staff and work-years	Estimated cost*	Status
5 COMPONENT: Develop draft watershed-based plan for the XX watershed. Commitment 5(a): Review guidelines and checklists for development of a watershed-based plan. Commitment 5(b): Meet with stakeholders in watershed. Commitment 5(c): Develop draft watershed-based plan. OUTCOMES/ENVIRONMENTAL RESULT: The watershed-based plan will allow for NPS problems and pollutants of concern to be addressed and reduced at a watershed level.	6/01 6/01 8/1	6/15 7/30 9/30	- Document activities in quarterly reports. - Document meetings in quarterly reports. - Submit draft watershed-based plan for EPA review.	WQS 0.14	$7,000	
6 COMPONENT: Provide training for NPS staff. Commitment 6(a): Attend the annual NPS Workshop Commitment 6(b): Attend workshops on BMPs and other nonpoint source issues as they become available. OUTCOMES/ENVIRONMENTAL RESULT: Staff knowledge will be increased, as demonstrated by staff through on-the-job implementation.	10/01	9/30	Provide status and summary of training in quarterly reports along with copies of training certificates, if available.	WQS 0.10 Crew	$5,000	
7 COMPONENT: Establish quarterly reporting to self-evaluate and joint-evaluate annual performance under the grant, including • A discussion of accomplishments as measured against the work plan • A discussion of the cumulative effectiveness of the work performed under all work plan components • A discussion of existing and potential problem areas • Suggestions for improvement, including, where feasible, schedules for making improvements. OUTCOMES/ENVIRONMENTAL RESULT: The tribe will evaluate and report on performance under the grant.	Quarterly: 1/30 4/30 7/30 10/30		Submit quarterly reports to EPA.	WQS 0.10	$2,000	

WQS: Water Quality Specialist at 1.00 FTE. Lead for all work plan components. Crew: NPS Crew. Contractual crew to carry out on-the-ground work for components #1–#3. EPA will have no role in carrying out the work plan commitments except to review quarterly reports submitted. * This reflects the totals of all COSTS for each component, including salary, fringe, equipment, contractual, supplies, travel, other direct costs (ODC), etc. Refer to budget breakdown for more detailed information.	TOTAL: $50,000 EPA: $30,000 MATCH: $20,000

Establishing Quality Assurance and Quality Control Protocols

In this section, you will learn
- *What EPA requires in terms of QA/QC*
- *Elements of a QAPP*
- *Links for additional information*

NPS pollution is usually addressed by collecting information on water quality, physical habitat, biological communities, land uses, land cover, and land management practices, and then using that information to select, size, design, install, operate, and maintain management practices that will intercept, treat, or otherwise reduce pollutant inputs and habitat degradation in a waterbody. Information quality is important because significant resources are often required to implement BMPs. Given that importance, tribes and other non-EPA organizations need to develop and implement quality systems to support their environmental programs and projects funded or regulated by EPA.

EPA has produced guidance on approaches for ensuring and controlling the quality of environmental data and project management. The guidance documents provide useful templates for establishing and maintaining QA/QC. Posted at *www.epa.gov/quality/ qa_docs.html*, they include program/organization tools, such as guidance for quality management plans, quality system audits, and QA training and reporting. In addition, there is guidance for projects such as producing data quality objectives, QA project plans, standard operating procedures, technical assessments, data validation and verification, and data quality assessments.

In general, EPA requires that recipients of funds for work involving environmental data collection comply with *Specifications and Guidelines for Quality Systems for Environmental Data Collection and Environmental Technology Programs* (ANSI/ASQC 2004). To demonstrate conformance, EPA requires two forms of documentation: (1) documentation of the organization's quality system (usually called a quality management plan) and (2) documentation of the application of quality-related activities to an activity-specific effort (usually called a quality assurance project plan or QAPP). For grants, contracts, and other agreements that consist of a single project or task, these two documents may be combined into a single document that describes the organization's quality system and the application of the system to the work performed under the grant or contract. This may be done only with the permission of the EPA QA manager, who will identify the elements that should be addressed in a combined document.

Tribes usually produce a QAPP that covers all aspects of project management and environmental data collection. The QAPP, which is usually developed and implemented to cover water quality monitoring and other activities supported by CWA section 106 funding, should be designed to address all data collection, including that funded by CWA section 319.

For example, a tribe with an approved QAPP for CWA section 106 monitoring can modify the QAPP so that it covers post-project monitoring of specific BMPs or the effects of several BMPs on water quality at a selected location downstream.

In general, the QAPP describes environmental data collection activities—through direct measurement or acquisition of databases—and explains who will be involved, what will be done, and how quality will be maintained. QAPPs (and modifications to QAPPs) must be prepared and approved before data collection begins. The QAPP will contain four basic sections:

- Project Management
- Data Generation and Acquisition
- Assessment and Oversight
- Data Validation and Usability

It will define and describe: who will use the data; what the project's goals/objectives or issues are; what decisions will be made from the information obtained; how, when, and where project information will be acquired or generated; what possible problems might arise and what actions can be taken to address them; the type, quantity, and quality of data desired; how *good* the data have to be to support the decisions to be made; and how the data will be analyzed, assessed, and reported. Figure I-13 depicts the role of the QAPP in data collection.

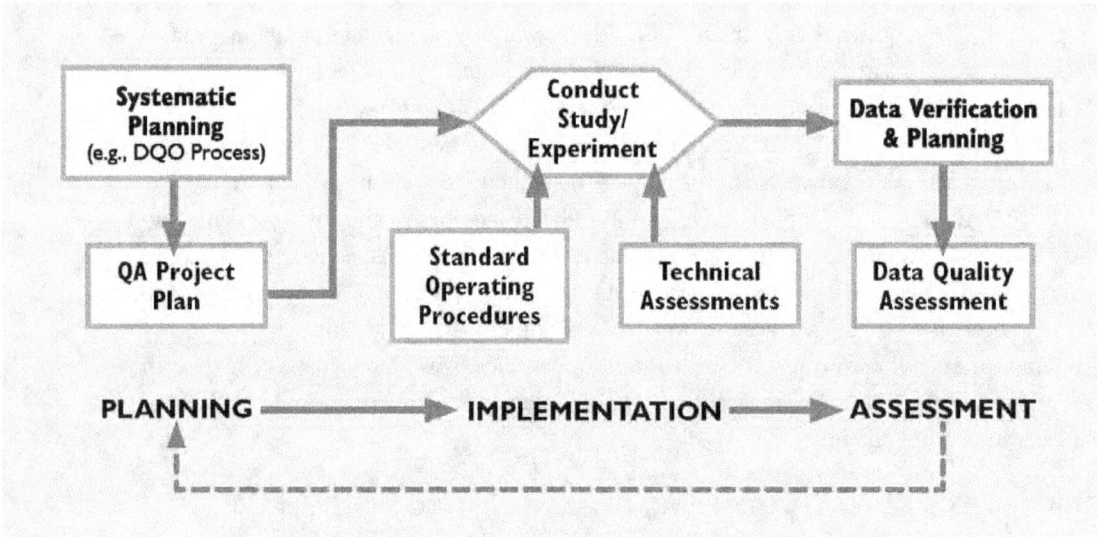

Figure I-13. The role of planning and QAPPs in data collection.

Tribes should ensure that data collection among CWA 106 and 319 programs is coordinated so that duplication of effort and overlap between programs are avoided. For example, a tribe can conduct all its data collection, including the CWA 319 efforts, under its CWA 106 program QAPP by modifying the QAPP to cover the NPS pollution activities. It will be important to ensure that the information collected applies to the specific objectives of the program or project. For example, collecting ambient samples to measure suspended sediment at strategic locations for the purpose of characterizing general water quality is different from collecting suspended sediment samples to measure the effects of a sediment-trapping BMP. The sampling locations, timing, and frequency will likely be different. These issues can be addressed in the QAPP by including separate sections for ambient monitoring and BMP performance monitoring.

EPA's *Guidance for Quality Assurance Projects Plans* is available at *www.epa.gov/quality/qs-docs/g5-final.pdf*.

Quality Assurance and Quality Control

Quality Assurance (QA): an integrated system of management activities involving planning, implementation, assessment, reporting, and improvement to ensure that a process, item, or service is of the type and quality needed. QA is typically applied by managers or technical personnel assigned to a specific oversight role. Example QA activities include technical and management assessments of field and analytical operations.

Quality Control (QC): an overall system of technical activities that measure the performance of a process, item, or service against defined standards to verify that the performance meets the stated requirements. QC is typically applied by technical personnel. Example QC activities include the use of control samples during sample collection, handling, and analysis, and activities such as data review.

Source: ANSI/ASQC 2004

Statutory Requirements of Section 319 Grants

In this section, you will find
- *The requirements to obtain 319 funds*
- *What reports are required by EPA*

The generally applicable award and administration process for assistance agreements funded under section 319 are governed by regulations at 40 CFR Part 31 (states, tribes, interstate agencies, intertribal consortia, and local governments). In addition, legal requirements, including EPA's regulations on environmental program grants for tribes (see 40 CFR 35.500 to 35.735) and regulations specific to NPS grants for tribes (see 40 CFR 35.630 to 35.638), apply to section 319 grants. A description of EPA's substantial involvement in the cooperative agreement must also be included in the final agreement. In addition, the following statutory requirements must be met to receive assistance funding.

Satisfactory Progress

For a tribe that received section 319 assistance agreements in the preceding fiscal year, section 319(h)(8) of the CWA requires that the EPA Region determine whether the tribe made *satisfactory progress* during the previous fiscal year in meeting the schedule of activities specified in its approved NPS management program to receive section 319 funding in the current fiscal year. The Region bases this determination on an examination of tribal activities,

Tour of a box timber weir installed at a culvert inlet to slow water before entering a rain garden. 2007 NPS Workshop, Region 5.

reports, reviews, and other documents and discussions with the tribe. Regions must include in each section 319 grant award (or a grant-issuance cover letter, signed by the EPA grant official) a written determination that the tribe has made satisfactory progress during the previous fiscal year in meeting the schedule of milestones in its NPS management program. Regional staff must include a brief explanation that supports the determination.

Administrative Costs

Pursuant to CWA section 319(h)(12), administrative costs in the form of salaries, overhead, or indirect costs for services provided and charged against activities and programs carried out with the grant may not exceed 10 percent of the grant award, regardless of whether the funding is base or competitive. The costs of implementing enforcement and regulatory activities, education, training, technical assistance, demonstration projects, and technology transfer are not subject to this limitation. It is common for work plans to include many of the above-stated exceptions to administrative costs. For example, most BMPs implemented by tribes are considered demonstration projects and would fall under the administrative cost exemption. Note that indirect cost *rates* are set by Department of Interior for the tribe and are independent of indirect costs mentioned in CWA.

Reporting

Performance and Financial Reports

All section 319 workplans must include a set of reporting requirements and a process for evaluating performance consistent with 40 CFR 31.40, 31.41, 35.507, 35.515, and 35.638. Tribes are required to submit performance reports and financial reports according to the schedule (at least annually, but no more than quarterly) determined by the EPA Regions.

Regardless of the funding distribution vehicle, tribes are responsible for managing the day-to-day operations and activities supported by the funding, ensuring compliance with federal requirements, and ensuring that milestones and performance goals are achieved. Tribes must submit performance reports and financial reports according to the schedule (at least annually, but no more than quarterly) in their assistance agreement. Copies of the performance reports are placed in the official files and provided to the recipient. Performance reports and financial

> *Water is the most critical resource issue of our lifetime and our children's lifetime. The health of our waters is the principal measure of how we live on the land.*
>
> —LUNA LEOPOLD

reports are due 30 days after the reporting period. The final report, sometimes called a closeout report, is due 90 days after the assistance agreement expires. Tribes are required to report direct and indirect environmental results from the work accomplished through the award.

Annual Reports

Section 319(h)(11) requires tribes to report annually on their progress in meeting the schedule of milestones contained in their NPS management programs. They must also report available information on reductions of NPS pollutant loadings (if available) and on improvements to water quality resulting from implementation of NPS management programs.

Annual reports should contain cumulative accomplishments, as well as how the management program goals are being met. Most tribes will not have information on load reduction estimates from NPS projects; however, staff should report on the following topics:

- A brief summary of progress toward meeting approved milestones and the goals and objectives identified in the NPS management program plan.

- A milestone matrix displaying information, such as description of project/program, scheduled project completion date, and percent completed.

- A description of how other federal, state, and tribal departments, community programs, and the like are supporting the NPS program.

- A summary of any water quality improvements or improvements to aquatic habitat as a result of the NPS program. Surrogate measures for water quality improvements, such as environmental indicators, may be used if water quality improvements are not yet available.

To simplify reporting, some tribes could choose to combine their last performance report with their annual report, as long as it also meets the requirements of the annual report and includes progress in meeting the schedule of milestones contained in their NPS management programs. Some tribes have chosen to include additional information in the annual report. For example, the report may discuss any needs to modify a program, provide case studies of particular projects, or convey information to a broader audience on the activities being conducted by the tribe. Tribes can choose to include this additional information as a way of marketing the successes of their program to tribal leaders and decision makers, or to the general public; this is, however, optional.

Summary and Conclusions

This section concludes **Part I** of the handbook. Although there are certainly many facets of the NPS program, some important take-away points include the following:

- Work with your EPA project officer or tribal NPS program coordinator (or both) throughout the eligibility process. They are responsible for reviewing the NPS assessment report and management program documents and for guiding you through the TAS process. Find out if there are Regional deadlines, in addition to national deadlines, to receive next year's fiscal year funding.

- Link work plan components to priorities in the NPS management program plan. All your NPS work and documentation—the assessment report, the management program plan, the 319 work plan, and so on—should "hang together" in an integrated manner and make sense to reviewers. Here is the key question regarding your assessment, program, and funding documentation: If you and all your staff left their jobs tomorrow and new people took over the program, could they pick up where you left off just by reviewing your work?

- Develop strong work plans that provide detailed information on each commitment and associated outputs and outcomes. Doing so will make tracking progress and reporting on environmental results easier for you and for EPA.

- It's a challenge for many small tribal environmental departments to maintain viable programs given the lack of resources and staff turnover. To the extent possible, partner with local, state, and federal agencies, as well as the nonprofit and private sectors, to leverage technical and financial assistance. Section 319 funding alone will not be able to solve all your NPS problems.

In September 2008, EPA's Office of Water released the *National Water Program Strategy: Response to Climate Change*. That document identifies potential effects of climate change on clean water and drinking water programs and defines actions that the National Water Program will take to address the effects. One of those actions is to integrate climate change information into existing training programs for water professionals. Given the magnitude and seriousness of the issue and the intent of the strategy document, EPA has included a special insert at the end of this part that addresses the effects of climate change on tribal NPS programs, including what managers and practitioners alike can do to incorporate climate change considerations into their programs and plans.

Part II of this handbook moves beyond the programmatic side of section 319 and provides more in-depth discussion of the technical aspects of improving water quality conditions through the use of watershed-based planning. Watershed-based planning incorporates all pollutant sources within a specific watershed geographic scale, rather than focusing on ownership-based boundaries. Unless a reservation is large enough to encompass entire watersheds, it is very likely that tribal water program staff will need to communicate with their neighbors to develop effective plans that cover entire watersheds.

Climate Change Considerations: The Earth Is Warming

Protection and restoration of tribal waters, watersheds, and the resources they support are increasingly important in light of changing climate. According to the Intergovernmental Panel on Climate Change (IPCC) 2007 report, the earth is warming. According to the global climate models, the warming will very likely cause changes in atmospheric circulation and increase evaporation and water vapor, which will result in precipitation increases, more intense precipitation, more storms, and sea level rise. The regional and local precipitation, temperature, and weather intensity are much harder to predict, but future projections suggest the following (IPCC 2008, 2007b; Burkett et al. 2001):

- Annual average precipitation will increase in the northeastern United States and decrease in the Southwest.
- In the Midwest and Great Lakes, lake and river levels will be lower.
- In the Great Plains, there will be intensified springtime floods and summertime droughts, and agricultural productivity will likely shift northward as the droughts increase.
- Projected warming in the western mountains by the mid-21st century is very likely to cause large decreases in snowpack, earlier snowmelt, more winter rain events, increased peak winter flows and flooding, and reduced summer flows.

Why does climate change matter to tribal NPS managers?

In addition to many cultural uses of the land and subsistence living, climate change is likely to affect many water-related issues depending on the location of your reservation. In general, warmer air temperature is expected to alter water in many ways.

- Changes in the location, timing, form, and amount of precipitation could result in
  - Reduced rainfall
  - More frequent wildfires (and land areas where wildfires have occurred are more vulnerable to soil erosion)
  - More frequent, more intense flooding

  - Increases in tropical storm intensity
- Hydrologic changes, presenting as
  - Shrinkage of the drainage network
  - Earlier peak runoff and lower summer flows in rain-snow watersheds
- Chemical and physical changes in oceans and coastal regions, including
  - Rising sea levels
  - Increasing erosion rate
  - Displacement of coastal wetlands
  - Inundation of coastal wetlands, deltas, and mangrove forests
  - Increases in the salinity of both surface water and ground water through saltwater intrusion
- Increases in water temperature, resulting in
  - Higher dissolved oxygen, pathogens, nutrients, ammonia, pentachlorophenol, and other pollutant levels
  - Increased algal blooms and invasive species
  - Loss of aquatic species whose survival and breeding are temperature-dependent
  - Change in the abundance and spatial distribution of coastal and marine species
  - Increased rates of evapotranspiration, shrinking waters such as lakes
- Increased evaporation from soils, leaving soils less able to support plant life and less able to absorb rain that does fall
- Protected wetlands and other water bodies may lose their relevance for species of concern because the protected areas will no longer provide the climate required

What should we be doing under the Tribal 319 Program?

Tribal NPS managers need to consider future threats as well as current sources of pollution when managing their resources, watersheds, and watershed processes. When you are developing your assessment report and management plans, developing watershed-based plans, and implementing BMPs, ask what would allow your practices and plans to be more resilient in the face of climate change. Below are just a few things you can do to adapt to and minimize the effects of climate change through your NPS program.

- When assessing and prescribing watershed management actions, project forward what the needs might be.
- Choose activities that protect or restore the resiliency of ecosystems and watersheds.
- Ensure your floodplains and hyporheic zones are maintained.
- When designing, selecting, and placing BMPs, consider that there might be more water flowing through the system during storms and less during summer. Fencing might need to be set back farther. Maybe headwater wetland restoration will take a higher priority than other projects so as to maximize storage from storm events.
- Do not assume that the protections and protected areas you have in place will be adequate for species as climate changes habitat.
- Plant trees. Retain and expand forests as much as possible; the extent of forests might shrink over time.
- Work with other tribal offices to consider climate change when permitting practices and developing codes/ordinances for forests, agricultural lands, developed areas, and water usage.

- Minimize increases in water temperature through shading and ground water recharge by protecting and restoring riparian areas and wetlands.
- Ensure that fish have access to seasonal habitat (e.g., off-channel and cool-water refugia).
- Disconnect impervious cover and road discharge from streams to soften discharge peaks during rain events.
- Use Green Infrastructure and Low-Impact Development. Green Infrastructure includes an array of products, technologies, and practices that use natural systems—or engineered systems that mimic natural processes—to enhance overall environmental quality and provide utility services. Green Infrastructure practices recharge ground water, reduce the need for watering vegetation, and reduce and slow excessive stormwater to streams.
- Reduce the reintroduction of carbon from stored carbon sources by minimizing soil and wetland disturbances and forest clearing.

Much of the information above came from EPA's climate change training materials, EPA Regional staff, *Ecological Impacts of Climate Change* (COEICC NRC 2008), and the University of Washington's Climate Impact Group. For the most current information about climate change research and adaptations and mitigation activities, see the climate change Web sites recommended in Part III.

PART II

Watershed-Based Planning

In this section of the handbook, we discuss additional resources for tribes that might be useful in the context of water resource management and program planning. The topics covered are

- Watershed approach and watershed-based plans (WBPs)
- Online tools
- Funding resources
- Environmental results
- Ideas for building partnerships to leverage resources and success

The Watershed Approach

The *watershed approach* is a holistic approach that can help tribes restore land and water resources. It is a framework that is particularly useful for preventing or correcting environmental problems caused by nonpoint source (NPS) pollution. It is important to distinguish between the watershed approach, which is the process or mechanism for planning, and the WBP, which is the action-oriented, end product of the planning process. This section outlines the six-step iterative watershed plan development process and the nine basic elements that are included in WBPs.

A few other distinctions should be made before launching into the steps of the watershed approach and the components of a WBP. There are many ways a WBP can be developed. Sometimes the plan is developed by a watershed group, university, local government, or nonprofit. For tribes, the tribal government might undertake the development of the plan directly or work in partnership with others. Tribal organizations bring a lot to the table during the plan development process, and they can work effectively with nontribal entities to create strategies to protect and restore the waters that flow through Indian Country and surrounding areas. Section 319 base or competitive funding can be used for best management practices (BMPs) and project work; in either case, the benefit to tribes is that a WBP plan builds the partnerships and collects the information critical to protecting and improving water resources.

The watershed approach has four characteristics:

1. Planning is conducted by a consortium of people working in diverse, well-integrated partnerships.

2. The focus for planning is a specific geographic area (the watershed).

3. Any actions that are included in the plan are based on sound science and technology.

4. The plan presents coordinated priority setting and integrated solutions.

It has been well documented that the watershed approach is the best means for preventing and resolving NPS problems and threats. The list below describes some of the benefits tribes will gain from using the watershed approach:

1. *It yields results.* Many people have achieved great results through this method, which endorses robust partnerships, a clear geographic focus, using sound science, and prioritizing issues.

2. *It helps prioritize.* What you submit in your grant application work plans will be derived from your WBP. The WBP is like a long-term improvement plan for the waters of the reservation, and adjacent waters if you are doing the plan based on a subbasin. The projects that you submit in annual work plans should be based on your NPS management plan document, but specific activities will come from the WBP as well.

3. *WBPs are supported by EPA and other federal agencies.* The U.S. Environmental Protection Agency (EPA) believes watershed-based planning is the most effective method to manage multiple water resource issues.

WBPs are like any other environmental project with up-front costs (time, people, and funding) that demonstrate results once implemented. WBPs are also beneficial in places where there is staff turnover—the next person can pick up the plan and read it, know who is who and what the long- and short-term goals and activities are, and be ready to go.

There is another distinction to be made—the difference between a *WBP* and the *NPS management program* required for 319 funding. Figure II-1 demonstrates how these documents work together. There are two main differences in the scope of the two documents:

1. The WBP focuses on a watershed, which is typically the drainage basin for a river or lake, and might be large-scale (8-digit Hydrologic Unit Code (HUC)) and contain areas outside tribal authority or smaller (16-digit HUC), which might be entirely on reservation lands. EPA's 319 program supports watershed planning at the 12-digit HUC or smaller level. However, the NPS management program focuses on all the waters of a reservation.

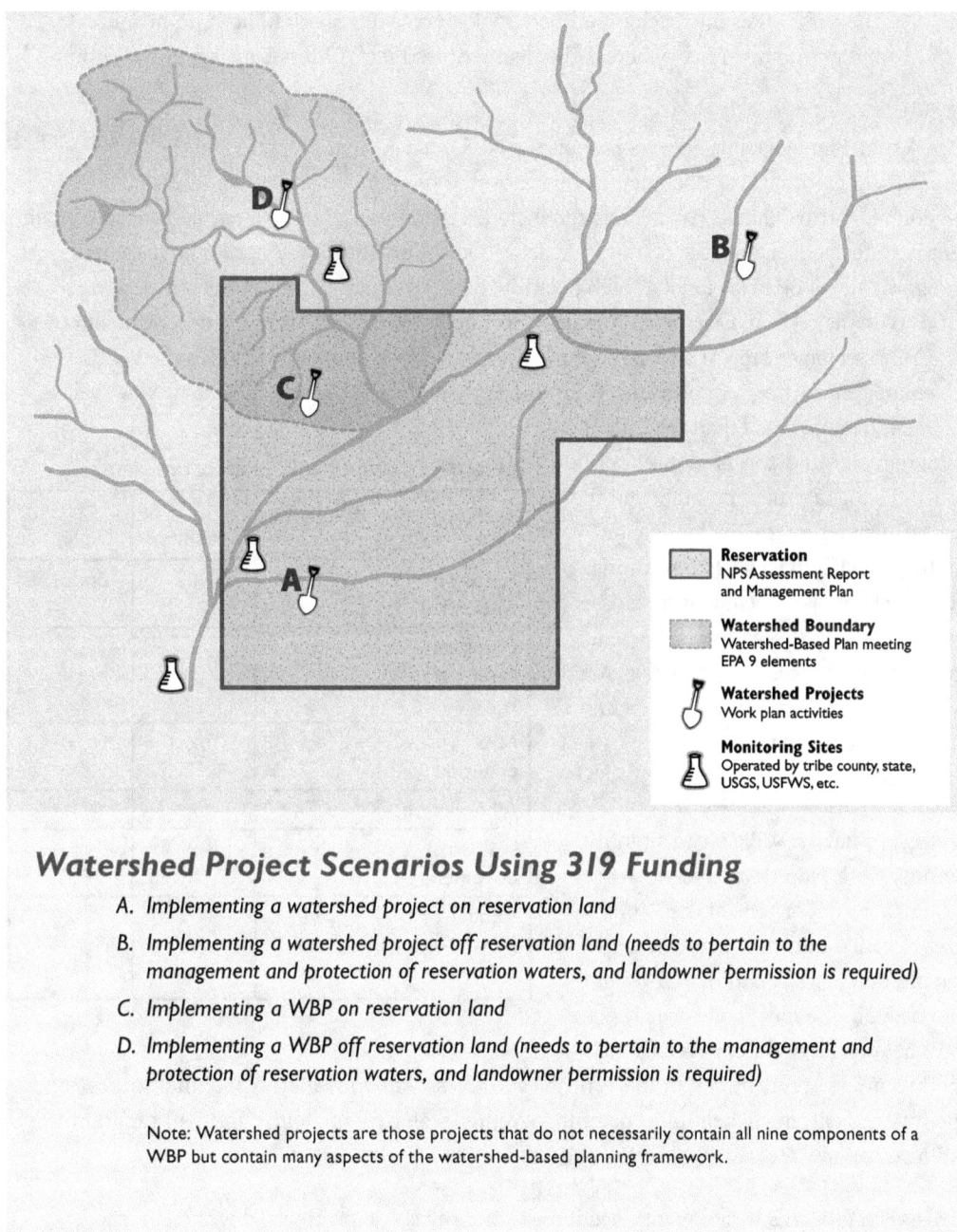

Watershed Project Scenarios Using 319 Funding

A. *Implementing a watershed project on reservation land*

B. *Implementing a watershed project off reservation land (needs to pertain to the management and protection of reservation waters, and landowner permission is required)*

C. *Implementing a WBP on reservation land*

D. *Implementing a WBP off reservation land (needs to pertain to the management and protection of reservation waters, and landowner permission is required)*

Note: Watershed projects are those projects that do not necessarily contain all nine components of a WBP but contain many aspects of the watershed-based planning framework.

Figure II-1. How a WBP relates to reservation-specific water resource plans and projects.

2. The WBP takes into account all potential sources of pollution, both point source and nonpoint source, whereas the plan required for 319 funds covers only nonpoint sources.

The foundation of a tribal NPS program is the NPS management program and the NPS assessment report. Those documents provide overall program guidance to address NPS pollution on tribal lands. Tribal lands can include several portions of watersheds or can contain entire watersheds depending on the scale definition. As an output of the NPS management program, the WBP helps to focus NPS planning on a particular watershed identified as a priority in the NPS management program. Information gathered during the development of the NPS assessment report and management program plan therefore feeds into the WBP. Like the management program plan, the WBP will be a multiyear planning document; however, it is much more detailed than the 319 program plan and covers a longer period. The NPS management program and the assessment report are the foundation to help you begin watershed planning, and the final plan document should be consistent with those reports. There are many other sources of information you will find helpful in writing your plan. You can use this handbook as a starting point to find these resources.

Once you have a WBP, it can simplify annual work plan development. Work plans identify portions of the WBP to be carried out. In the case of work plans submitted as part of a 319 grant application, the focal point would still be on NPS pollution even though point sources are identified in the WBP. Nevertheless, knowing all the potential sources of pollution is extremely helpful in planning a robust water resource management program for all reservation waters.

Table II-1. Components of assorted plans

Component	NPS mgmt plan	WBP	Work plan
Focus on watershed	optional	x	optional
Focus on reservation waters	x		x
NPS pollution	x	x	x
All pollution		x	
Multiyear document	x	x	
Annual (1- to 2-year) document			x

This section begins by going into detail regarding the six steps involved with watershed planning (i.e., the *approach*). It then takes you through each of the nine required components of the WBP (highlighted in white in Figure II-2). To understand how these fit together, see Figure II-2. Each of the nine components required in a WBP is derived from one of the six planning and implementation steps. This section also takes a look at the many resources available to help you develop a watershed plan.

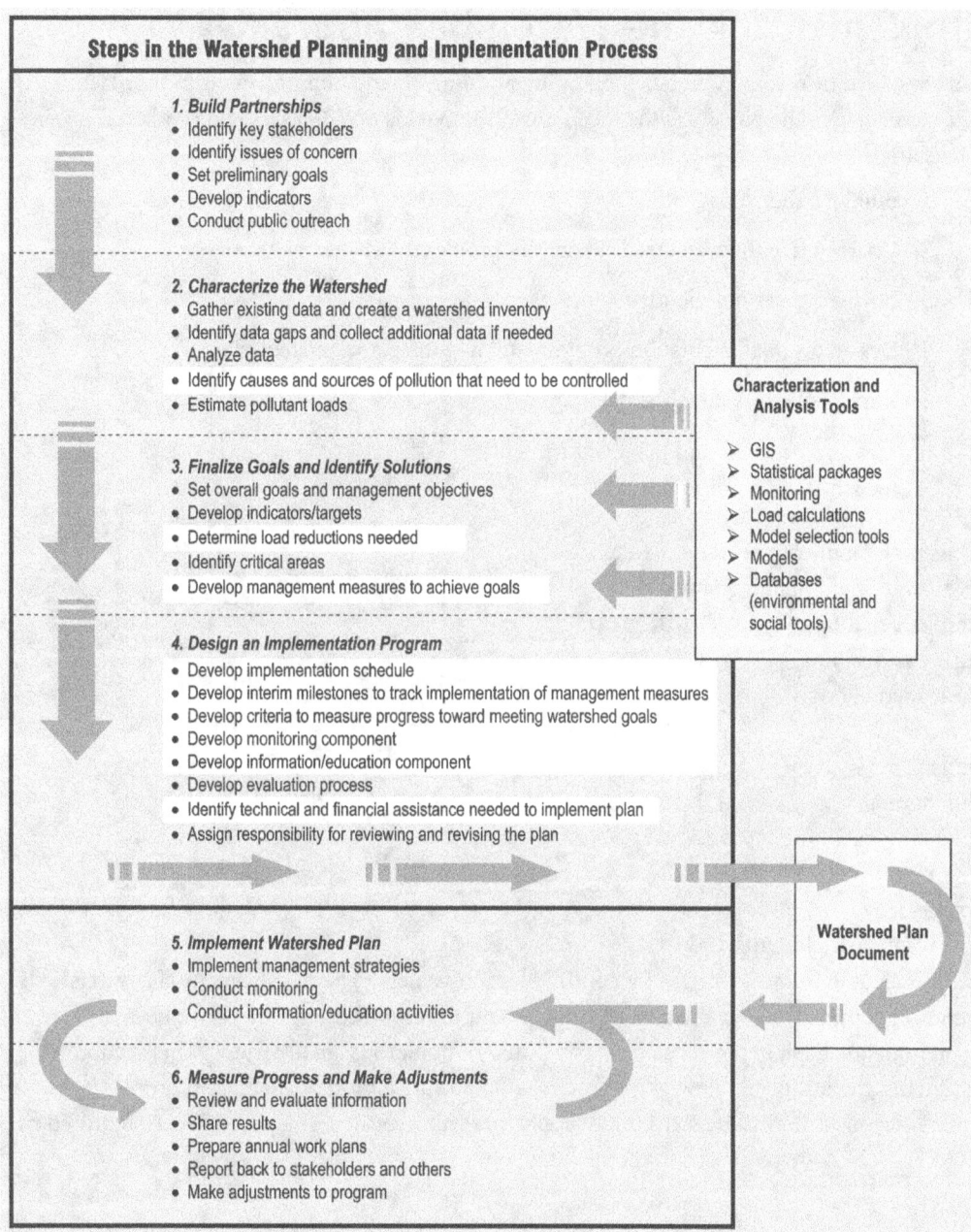

Figure II-2. Foundation of the nine elements of a WBP.

Implementing the Watershed Approach

As explained below, the watershed planning and implementation process (the approach) has six major steps. The process is the mechanism that yields a WBP as its end product. The six steps are:

1. Build partnerships

2. Characterize the watershed to identify problems

3. Finalize goals and identify solutions

4. Design an implementation program and assemble a watershed plan

5. Implement the watershed plan

6. Measure progress and make adjustments (which lead to an improved plan)

This diagram illustrates the six steps associated with watershed planning and implementation. Note that the diagram is shaped like a feedback loop. This is intentional: like any good management process, the watershed planning and implementation process strives for continual improvement. The management of a watershed requires continual self-evaluation and adjustment to reach the goals that have been set for water quality. Each of the six steps is composed of numerous smaller steps. This section discusses the utility and flow of the watershed planning and implementation process, with an emphasis on how these steps might apply in a tribal setting. The *nine elements*, required by EPA for WBPs, are discussed in the next section.

I. Build Partnerships

Bringing people, policies, priorities, and resources together through a watershed approach blends science and regulatory responsibilities with social and economic considerations. Because building partnerships is the first step and a critical one for ensuring the ongoing success of your efforts, EPA has included in this section a good amount of detail regarding identifying stakeholders and nurturing those relationships.

Watershed planning is often too complex and too expensive for one person or organization to tackle alone. To work effectively at the watershed level (as opposed to limiting the efforts to only tribal lands), tribes will want to work with stakeholders within and outside the tribe and watershed. A *stakeholder* is a person or organization that has a stake in the outcome of the watershed planning process—a stakeholder can make and implement decisions, is affected by the decisions made, or has the ability to assist or impede implementation of the decisions. Figure II-3 offers a starting point for thinking about which tribal and nontribal stakeholder groups to include in your consortium. Note that key individuals might be the impetus within each stakeholder group.

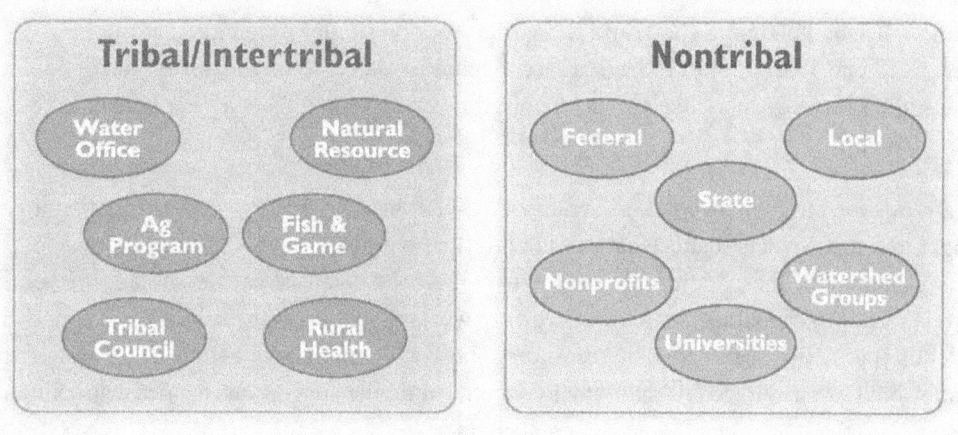

Figure II-3. Potential stakeholder groups for your consortium.

A tribe might be faced with the issue that its reservation constitutes only a very small portion of the watershed; land titling issues might be complex; or lands might be "checker-boarded." In all such cases, land-based activities occurring on nontribal lands are likely to account for a portion—and perhaps even the majority—of the NPS pollution affecting tribal lands. For example, tribal lands might be downstream of agricultural operations or heavily paved urban areas. This is why partnering with nontribal entities is so important. Tribes might not be able to solve their water pollution problems without the cooperation of their neighbors in the watershed. By the same token, the nontribal groups will likely need tribal involvement to address the water quality problems they're facing.

Note that tribal 319 funding generally applies to waters on the reservation, though the watershed might cover a much larger geographical area. If your work plan tasks include implementing 319 projects outside tribal reservations, you will need to make the case that these projects in upstream or downstream waters pertain to the management and protection of reservation waters. In addition, approval from those property owners will be needed to implement such projects off the reservation.

When Should We Start Engaging Stakeholders?

It is critical to build partnerships with key interested parties *at the outset* of the watershed planning effort. A common mistake is engaging stakeholders after much of the planning has already been done. That can lead to disagreements or lack of enthusiasm, which might have been avoided if stakeholders had been brought on board earlier.

Who Should We Include?

All categories of potential stakeholders—not just those that volunteer to participate—should be identified and invited. It is essential that you identify all these categories of potential stakeholders. Stakeholders also include those that can contribute resources and assistance to the watershed planning effort, and those that work on similar programs that can be integrated into a larger effort or have access to lands or waters to be addressed. Keep in mind that stakeholders are more likely to get involved if you can show them a clear benefit to their participation.

Begin by contacting the people and organizations that have an interest in water quality or might become partners that can assist you with the watershed planning process. Consider who would be the most appropriate person to contact the potential partner. Make sure you have a means to encourage the partners you are working with to invite others. This approach extends your network and possible resources. Those who might have a stake in the watershed plan should be encouraged to share their concerns and offer suggestions for possible solutions. Also try to match your needs with your stakeholders' resources and capabilities. The box below lists some skills you might seek to complement tribal resources and knowledge.

Skills in Stakeholder Group	Resources Available from Stakeholders
■ Accounting	■ Contacts with media
■ Graphic design	■ Access to volunteers
■ Computer support	■ Access to datasets
■ Fundraising	■ Connections to local organizations
■ Public relations	■ Access to meeting facilities
■ Technical expertise (e.g., GIS, water sampling)	■ Access to equipment
■ Facilitation	■ Access to field trip locations
■ Consensus-building, outreach	■ Access to upstream locations

Remember that partners should come from both tribal/intertribal entities and nontribal entities. Other agencies within tribes might be able to help tap into outreach demographics previously untargeted by tribal water quality efforts. Working with tribal leaders can lead to swifter governing policies that support the water quality plan. Nontribal entities provide access to outside resources that bolster research and legislative efforts. A good place to start when thinking about possible partners is making a list of the expertise and resources that will be needed to ensure success. Needs range from technical data to meeting space for gatherings. A list will help target your partnership efforts.

How Do We Build Support in the Watershed?

Watershed plans and water quality management programs also contribute to stakeholders by addressing their needs and concerns. Listing issues relevant to the local community will help with buy-in from tribal members and organizations. Restoring clean water and preserving natural resources are ways that the tribe can benefit from a WBP. Restored areas might be less prone to flash flooding during large storm events. Encouraging responsible development can help spur economic growth, yet minimize contribution to the degradation of reservation waters. Water quality management efforts, monitoring sites, and BMP installations also preserve the land for tourism and may be marketed to *green* tourists interested in environmental protection.

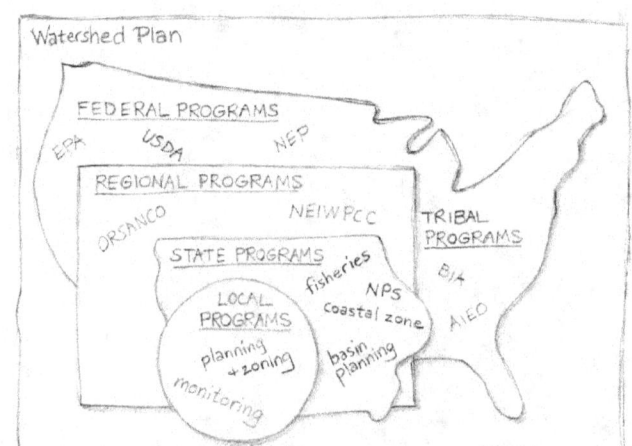

Figure II-4. Hierarchy of entities within a watershed.

Very few watersheds exist within a single political jurisdiction. Many different regulatory agencies are involved in any watershed plan simply because of the large area affected and the cross-cutting nature of water quality issues—and those agencies do not always agree on the best course of action. Creating partnerships with state and federal agencies can facilitate consensus in your watershed planning. When all groups are on the same page about what a watershed needs, where funding will come from, who will maintain any infrastructure or testing protocol, and what the benefits will be, projects are approved more quickly and run more smoothly. Without reaching out to partners, the effort to improve reservation waters can seem like an uphill battle. In actuality, water quality is important to everyone and good communication can make that clear.

When you begin to contact stakeholders, consider how you will approach them about your watershed planning effort and the expected outcome. Think about what will encourage that stakeholder to buy in to your process, and focus your approach on issues important to that stakeholder or group. For community organizations, you can cite the resolution of problems like erosion of stream banks, habitat and *green space* deterioration, and flash flooding.

Amber Marriott

Policy organizations will be focused on what your watershed plan can do for the legislative process: Will it help streamline environmental policy? Does it contribute data required by local or state ordinances, or does it eliminate the need for deliberations on community issues because it addresses these issues? Try to contact the most appropriate person in the organization (based on how you approach that stakeholder's buy-in to the project). Water quality contacts are naturally good contacts, as are environmental committee members, school excursion leaders, or community government officials.

Which Stakeholders Might Be Able to Assist Monetarily?

Funding is one of the most important parts of the watershed planning process. Many federal programs have funding opportunities that can be used in your watershed plan. Look for funding in these topics: wetlands restoration, water quality restoration, community development and aid, historical preservation, solid waste cleanup, land reclamation and revegetation, abandoned mine reclamation, and other programs targeted toward the needs in your watershed. When contacting federal and state agencies, it is best to research your local branch. Bureau of Indian Affairs, Bureau of Land Management, U.S. Department of Agriculture (USDA)–Natural Resources Conservation Service (NRCS), U.S. Geological Survey, and EPA have information on local representatives on their respective Web sites. Getting in contact with these local branches can have positive results because they can be untapped resources. In addition, if you are working with local universities, encourage your partners to apply for federal research grants that will support their work on your watershed plan. For links to online federal funding resources, see **Part III**, Additional Resources for Tribes.

Included below is a list of example stakeholders in the watershed planning process. A more comprehensive list is provided in EPA's *Handbook for Developing Watershed Plans to Restore and Protect Our Waters* (*www.epa.gov/nps/watershed_handbook*), Chapter 3.

Local Stakeholders	State and Regional Programs	Federal Programs and Organizations
■ *Community service organizations*	■ *Source Water Assessment and Protection (SWAP) programs*	■ *Abandoned mines programs*
■ *Local cooperative extension offices*	■ *State and interstate water commissions*	■ *Agricultural conservation programs*
■ *Local elected officials*	■ *State coastal zone management programs*	■ *Agricultural support programs*
■ *Local lake associations*	■ *State departments of transportation*	■ *Bureau of Indian Affairs*
■ *Local landowners*	■ *State fish and wildlife programs*	■ *Bureau of Land Management*
■ *Parks and recreation departments*	■ *State health departments*	■ *Coastal programs*
■ *Planning and zoning programs*	■ *State TMDL (total maximum daily load) programs*	■ *Federal transportation programs*
■ *Regional planning councils*	■ *State NPS programs*	■ *Indian Health Service*
■ *Soil and water conservation districts and NRCS offices*	■ *State water protection initiatives*	■ *NRCS*
■ *Solid waste programs*	■ *State wetland programs*	■ *Public lands management*
■ *Stormwater management programs*	■ *Regional geographic watershed initiatives*	■ *Threatened and Endangered Species Protection Programs*
■ *Volunteer monitoring programs*		■ *U.S. Fish and Wildlife Service*
■ *Water and sewer programs*		■ *U.S. Forest Service*
■ *Watershed organizations*		■ *Wetland protection programs*
		■ *Wildlife protection programs*

2. Characterize Your Watershed

Characterizing your watershed can be considered the heart of the preparation process for WBPs. The process is very similar to producing the NPS assessment report for the tribal Clean Water Act (CWA) 319 program, but it also includes point sources of pollution, such as wastewater treatment plants, industrial facilities, and stormwater discharges permitted under the National Pollutant Discharge Elimination System (NPDES). During this phase, you will gather existing data and create a data inventory, analyze gaps and collect additional data if needed, and analyze your data. The goal of this data analysis for the watershed is to identify the *causes and sources* of pollution affecting your waterbody. At the end of this characterization process, you should be able to state what kinds of pollution are affecting the waterbody, where that pollution is coming from, and what you might not know about the pollutant source.

The first step of watershed characterization is determining the scope of your effort, in terms of both geographic size and the number of pollutant causes and sources you will address.

Areas are defined by two-digit codes, so watersheds will be at the 2-, 4-, 6-, 8-, 10-, and 12-digit level. An easy way to remember the relationship between HUCs and the land they represent is this: The smaller the number, the larger the geographic area. For that reason, effective plans tend to focus at the 12-digit HUC (or smaller), level. The 12-digit level, or *subwatershed*, usually has 10,000 to 40,000 acres; some are as small as 3,000 acres. EPA encourages plans at the 12-digit level or smaller.

The key to determining the appropriate scale of planning is that you must ensure that the area is small enough to manage but large enough to address water quality impairments and the concerns of stakeholders. If the scale is too small, significant sources of pollution that are outside the planning area might be mistakenly ignored, and then you cannot fully address the problem. If the scale is too large, you might be overwhelmed. Remember that your scope is both geographic and topical—where you want to work and on what parameters. The greater number of pollutants, the smaller the geographic scale and vice versa.

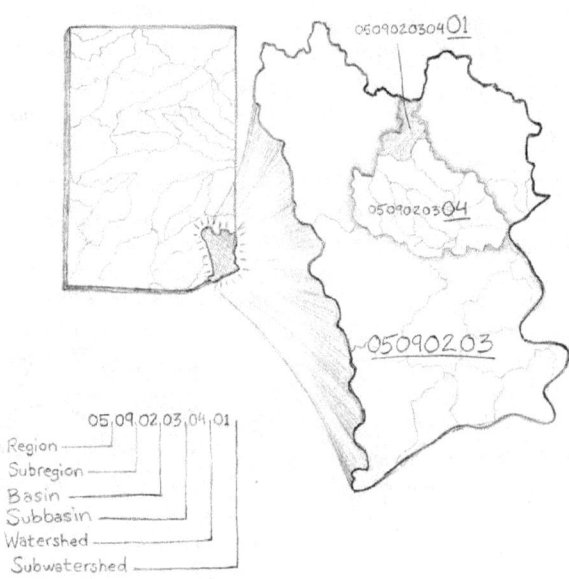

Figure II-5. HUC code levels used to define watershed size.

Once you have delineated the area within which you will work, set out to gather as much relevant information on that area as possible, with the end goal of identifying the causes and sources of pollution. In addition to meeting with stakeholders, you will want to thoroughly review the data that have already been collected. Often, likely causes of pollution have already been identified. For example, stakeholders might be aware of a large construction project, paved roads, or extensive agricultural operations near their waterbody. Table II-2 lists the pollutants that might correspond with those sources. The process of characterizing your watershed leads from generalities to specifics, so tailor the identification to what is applicable in your delineation. Your chart will most likely differ from that in Table II-2.

Table II-2. Impairment sources and associated pollutants

Source	Common associated pollutants
Cropland	Turbidity, phosphorus, nitrates, temperature, total suspended solids
Forestry harvest	Turbidity, temperature, total suspended solids
Grazing land	Fecal bacteria, turbidity, phosphorus, nitrates, temperature
Industrial discharge	Temperature, conductivity, total solids, toxic substances, pH
Mining	pH, alkalinity, total dissolved solids, metals
Septic systems	Fecal bacteria (e.g., *Escherichia coli*, enterococci), nitrates, phosphorus, dissolved oxygen/biochemical oxygen demand, conductivity, temperature
Sewage treatment plants	Dissolved oxygen and biochemical oxygen demand, turbidity, conductivity, phosphorus, nitrates, fecal bacteria, temperature, total solids, pH
Construction	Turbidity, temperature, dissolved oxygen and biochemical oxygen demand, total suspended solids, toxic substances
Urban runoff	Turbidity, total suspended solids, phosphorus, nitrates, temperature, conductivity, dissolved oxygen and biochemical oxygen demand

Types of Data to Review to Characterize Your Watershed

Physical and Natural Features

- *Watershed boundaries*
- *Hydrology*
- *Topography*
- *Soils*
- *Climate*
- *Habitat*
- *Wildlife*

Land Use and Population Characteristics

- *Land use and land cover*
- *Existing management practices*
- *Demographics*

Waterbody Conditions

- *Water quality standards/ designated uses*
- *305(b) report*
- *303(d) list*
- *Integrated report*
- *TMDL reports*
- *Source Water Protection Areas*

Pollutant Sources

- *Point sources*
- *Nonpoint sources*

Waterbody Monitoring Data

- *Water quality data*
- *Flow data*
- *Biological data*

Where might you find these existing data? A number of entities regularly collect all kinds of information in your watershed (see list below). A good place to start is with federal and state agencies, as well as any documents prepared during your 106 and 319 application processes. For example, you have probably uploaded data collected for your 106 reports into EPA's STORET database.

- Federal agencies (U.S. Geological Survey [USGS], U.S. Fish and Wildlife Services, U.S. Forestry Service, Bureau of Land Management, U.S. Army Corps of Engineers, EPA, EPA/STORET)

- State agencies (water, fish and game, forestry, agriculture)

- Colleges and universities

- Watershed groups (volunteer monitoring programs, local knowledge)

- Lake and river associations

- Local agencies (water/wastewater, health, planning and zoning, and the like)

- Regional planning agencies

- EPA's *Surf Your Watershed* at *www.epa.gov/surf*

If you do not already have a data inventory, create one so that all relevant information can be easily accessed and revisited in the future. The types of information to include in your inventory are:

- Type of data (e.g., monitored, geographic)

- Source of data (agency)

- Quality of data (quality assurance/quality control documentation, quality assurance project plan [QAPP])

- Representativeness of data (number of samples)

- Spatial coverage (location of data collection)

- Temporal coverage (period of record)

- Data gaps

You can find sample inventories in EPA's *Handbook for Developing Watershed Plans to Restore and Protect Our Waters*. Once you have created the data inventory, you will move on to the next phase in characterization: *identify gaps and collect new data*. As you review the data, you might realize that you need to gather additional existing information. If so, go back, add additional information to your data inventory, and then proceed, filling any gaps in your data. You will know that you have collected enough data once you are able to (1) identify the causes and sources of the pollution in your waterbody of concern and (2) verify that the quality of your data is adequate for making those determinations.

A watershed survey, or visual assessment, is one of the most rewarding and least costly means of collecting the additional data needed to understand the pollutants of concern in the watershed. By walking, driving, or boating the watershed, you can observe water and land conditions, uses, and changes over time that might otherwise be unidentifiable. These surveys can help you identify and verify pollutants, sources, and causes, such as streambank erosion delivering sediments into the stream and illegal pipe outfalls discharging various pollutants. They can also be used to familiarize local stakeholders, decision makers, citizens, and agency personnel with activities occurring in their watershed. Additional monitoring of chemical, physical, and biological conditions will be required to determine whether the pollutants observed are actually affecting the water quality. In addition, aerial photos are an excellent resource for viewing NPS impacts. For general information on visual surveys, read section 3.2, The Visual Assessment, in EPA's *Volunteer Stream Monitoring: A Methods Manual* (EPA 841-B-97-003), *www.epa.gov/owow/ monitoring/volunteer/stream/vms32.html.* Included is a Watershed Survey Visual Assessment form, *www.epa.gov/owow/monitoring/volunteer/stream/ds3.pdf.*

Several agencies and organizations have developed visual assessment protocols that you can adapt to your own situation. For example, NRCS has developed a Visual Stream Assessment Protocol (VSAP). The VSAP is an easy-to-use assessment tool that evaluates the condition of stream ecosystems. Go to *www.nrcs.usda.gov/technical/ECS/aquatic/svapfnl.pdf* to download a copy of the tool.

For more information and detailed descriptions of water quality sampling methods and sampling plan designs, see the USGS *National Field Manual for the Collection of Water-Quality Data* at *water.usgs.gov/owq/FieldManual* and EPA methods at *www.epa.gov/quality/qs-docs/g5s-final.pdf.*

At this point you have inventoried and evaluated existing information and collected new data if necessary, partnering with other entities to the extent possible to get the information you need. The outcome of this process for tribes is to "identify causes and sources of pollution." This fulfills the first of the nine elements discussed in more detail in the subsequent section called Nine Elements of a Watershed-Based Plan. At this point you should be able to characterize your causes and sources on the basis of the following:

- Source type (e.g., nonpoint, point)
- Location (e.g., subwatershed)
- Land use type
- Source behavior (e.g., direct discharge, runoff, seasonal activities)

Tribes also have the option at this stage of estimating the quantities (loads) of the pollutants of concern entering their waterways. Two general types of techniques for estimating pollutant loads can be used. First, you can directly estimate loads from monitoring data or literature values. Such techniques are best suited to conditions where fairly detailed monitoring and flow gauging are available, and the major interest is in total loads from a watershed. To obtain literature values for estimating pollution loads, see EPA's *Handbook for Developing Watershed Plans to Restore and Protect Our Waters*, figure 8.1. Second, where resources allow, you can select from a number of different watershed modeling techniques. Models can be used to forecast or estimate future conditions; many do, however, require extensive expertise. One simple model that might be of help is EPA's *Spreadsheet Tool for Estimating Pollutant Loads* (STEPL). If the pollutants of concern are sediment, nitrogen, or phosphorus, you can use EPA's STEPL spreadsheet tool to estimate loads and load reductions. The model uses simple algorithms to calculate nutrient and pollutant sediment loads from different land uses and the load reductions that would result from implementing various BMPs. To download the model or read its documentation, visit *http://it.tetratech-ffx.com/stepl*.

3. Finalize Goals and Identify Solutions

During this part of the planning process you will: set goals on the basis of your data analysis; identify your management objectives; and select your indicators to measure progress toward achieving water quality improvements. Now that you have characterized and quantified the problems in the watershed, you are ready to refine the goals and establish more detailed objectives and targets that will guide you in developing and implementing a management strategy. Those goals will guide the identification and selection of management practices to meet the targets and, therefore, the overall watershed goals.

Table II-3 provides some examples of translating watershed goals into management objectives (From *EPA's Handbook for Developing Watershed Plans to Restore and Protect Our Waters*, chapter 9).

The *indicators* are measurable parameters that will be used to link pollutant sources to environmental conditions. The specific indicators will vary depending on the designated use of the waterbody (e.g., warm-water fishery, cold-water fishery, recreation) and the water quality impairment or problem of concern. For example, multiple factors might cause degradation of a warm-water fishery. Potential causes include changes in hydrology, elevated nutrient concentrations, elevated sediment, and higher summer temperatures. Each of these stressors can be measured using indicators like peak flow, flow volume, nutrient concentration or load, sediment concentration or load, and temperature.

A specific value can be set as a target for each indicator to represent the desired conditions that will meet the watershed goals and management objectives. Targets can be based on water quality criteria or, where numeric water quality criteria do not exist, on data analysis, reference conditions, literature values, or expert examination of water quality conditions to identify values representative of conditions that support designated uses. If a total maximum daily load (TMDL) already exists for pollutants of concern in your watershed, you should

Table II-3. Translating watershed goals into management objectives

Preliminary goal	Indicators	Cause or source of impact	Management objective
Support designated uses for aquatic life; reduce fish kills	Dissolved oxygen Phosphorus Temperature	Elevated phosphorus causing increased algal growth and decreased dissolved oxygen Cropland runoff	Reduce phosphorus loads from cropland runoff and fertilizer application
Reduce flood levels that might cause water quality problems	Peak flow volume and velocity	Inadequate stormwater controls, inadequate road culverts	Minimize flooding impacts by improving peak and volume controls on urban sources and retrofitting inadequate road culverts
Restore aquatic habitat	Riffle-to-pool ratio, percent fine sediment	Upland sediment erosion and delivery, streambank erosion, near-stream land disturbance (e.g., livestock, construction)	Reduce sediment loads from upland sources; improve riparian vegetation and limit livestock access to stabilize streambanks
Improve aesthetics of lake to restore recreational use	Algal growth, chlorophyll *a*	Elevated nitrogen causing increased algal growth	Reduce nitrogen loads to limit algal growth
Restore wetland	Populations of wetland-dependant plant and animal species; nitrogen and phosphorus	Degradation of wetland causing reduced wildlife and plant diversity and increases in nitrogen and phosphorus runoff because of a lack of wetland filtration	Restore wetland to predevelopment function to improve habitat and increase filtration of runoff
Conserve and protect critical habitat	Connectivity, areal extent, patch size, population health	Potential impacts could include loss of habitat, changes in diversity, etc.	Maintain or improve critical habitat through conservation easements and other land protection measures

review the TMDL to identify appropriate numeric targets. TMDLs are developed to meet water quality standards, and when numeric criteria are not available, narrative criteria (e.g., prohibiting excess nutrients) must be used to develop numeric targets.

Because tribes are not required to base improvements on load reductions, they have the discretion to set a wide range of water quality-based goals. Examples include the following:

- Meet state or tribal water quality standards for one or more pollutants or uses

- Improve measurable water quality conditions or parameters, such as in-stream reductions in a pollutant or improvements in a parameter that indicates stream health (e.g., macroinvertebrate counts)

- Enhance/restore fisheries

- Stabilize stream banks

- Restore ceremonial waters

Once you have selected your water quality-based goals, you can begin to think about the techniques, also called BMPs, to meet those goals. You can find in-depth information on various BMPs at *www.waterquality.utah.gov/TMDL/Virgin_River_Watershed_Implementation_Appendix.pdf.*

Given the wide range of BMP choices, how do you know which ones will be right for you? Which will give you the results you need and also be acceptable to stakeholders and reasonably affordable? Look to see what people are already using in your area that is working, and then determine whether something different is needed. One useful tool is the BMP effectiveness tables in the document *National BMPs to Control Nonpoint Pollution from Agriculture* available at *www.epa.gov/nps/agmm.* The table on page 16 of that document lists various agricultural management practices in the left column. The other columns provide some relative effectiveness information, for each BMP, for controlling runoff volume, phosphorus, nitrogen, sediment, and bacteria. Tables like that one, which aggregate BMP performance, can be used to screen various management practices and provide information on what might work in a particular situation to reduce the effects of NPS pollution.

4. Design an Implementation Program

Now that you have identified watershed BMPs that when implemented should meet your objectives, it is time to develop the remaining elements of your implementation program. Designing the implementation program generates several of the basic elements needed for effective watershed plans:

- An information/education component to support public participation and build management capacity related to adopted BMPs

- A schedule for implementing BMPs

- Interim milestones to determine whether BMPs are being effectively implemented

- Criteria by which to measure progress toward meeting water quality-based goals

- A monitoring component to evaluate the effectiveness of implementation efforts

- An estimate of the technical and financial resources and authorities needed to implement the plan from both your organization and partners

- An evaluation framework

Thus, the end of this process of information gathering and analysis should yield a WBP that you can use as a tool for the next 5 or so years. The required elements of this plan are discussed in detail in the next section.

5. Implement Your Plan

Although many watershed planning handbooks end with development of the plan, the plan is just the starting point. The next step is to implement the plan in your watershed. Implementation can begin with an information/education component or with on-the-ground BMPs and projects that implement BMPs. The plan should prioritize which steps are the most important so you will know where to begin.

When implementation begins, the dynamic of your watershed partnership, and stakeholders' level of participation, might change. After the plan is completed, you need to determine how you want to continue to operate. Implementing a watershed plan involves a variety of expertise and skills, including project management, technical expertise, group facilitation, data analysis, communication, and public relations. Your watershed plan implementation team should include members who can bring such skills to the table. The BMPs you selected, the schedules and milestones you set, the financial and technical resources you identified, and the information/ education programs you developed in the course of assembling your plan provide a road map for implementation. Follow it. Take advantage of the partnerships you formed during plan development to work toward efficient plan implementation.

Key implementation activities include the following:

- Ensuring technical assistance in designing and installing BMPs
- Providing training and follow-up support to landowners and other responsible parties in operating and maintaining the BMPs
- Managing the funding mechanisms and tracking expenditures for each action and for the project as a whole
- Conducting land treatment and water quality monitoring activities and interpreting and reporting data
- Measuring progress against schedules and milestones
- Communicating status and results to stakeholders and the public
- Coordinating implementation activities among stakeholders, among multiple jurisdictions, and within the implementation team

To keep the implementation team energized, consider periodic field trips and site visits to document implementation activities in addition to the necessary regular team meetings.

Other tasks associated with implementing the watershed plan include preparing work plans. These plans will outline the implementation activities in annual or even 2- to 3-year time frames. Think of your watershed plan as a strategic plan for long-term success. Annual grant work plans are the specific to-do lists to achieve the vision expressed in your watershed plan.

Think of your watershed plan as a strategic plan for long-term success. Annual grant work plans are the specific to-do lists to achieve the vision expressed in your watershed plan.

As you work through various tasks, share your results. Continuous communication is essential to building the credibility of and support for the watershed implementation process. Lack of communication can impede participation and reduce the likelihood of successful implementation. This is especially critical if you are using a stakeholder-driven process. Transparency of the process builds trust and confidence in the outcome. Regular communication also helps to strengthen accountability among watershed partners by keeping them actively engaged.

Progress and implementation results can be shared through various media formats, such as press releases, ads in local newspapers, television or radio public service announcements, or presentations at community meetings such as those of homeowner associations and local civic organizations, Parent Teacher Association meetings, or other gatherings of members of the watershed community. You could secure time on the local cable access station to discuss the watershed plan and share monitoring results with the public. You might also consider hosting a press conference with local officials and the stakeholders as a way to thank them for their participation, and to inform the larger community about the plan's contents and how they can participate in implementing the plan.

6. Measure Progress and Make Adjustments

Remember to publicize the project team's accomplishments to tribal councils or decision makers, county commissioners, elected local and state officials, funders, watershed residents, and other stakeholders. In **Part I**, you learned that 319 annual reports may include a list of accomplishments to share with interested stakeholders. The project team might also wish to issue a watershed *report card* or develop a fact sheet or brochure to highlight its successes. Report cards let the community know whether water quality conditions are improving overall. They also allow people to compare results across specific areas to see if things are improving, whether some aspects seem to be connected, and whether a change in direction is needed to bring about greater improvements. This is an effective way to build awareness of the watershed issues and the progress of watershed plan implementation. In addition, when people see progress, they will continue to work toward making the plan a success.

Figure II-6 demonstrates how the watershed approach is an iterative process with continual improvement as its main goal (adapted from EPA's *Handbook for Developing Watershed Plans to Restore and Protect Our Waters,* diagram 13-2).

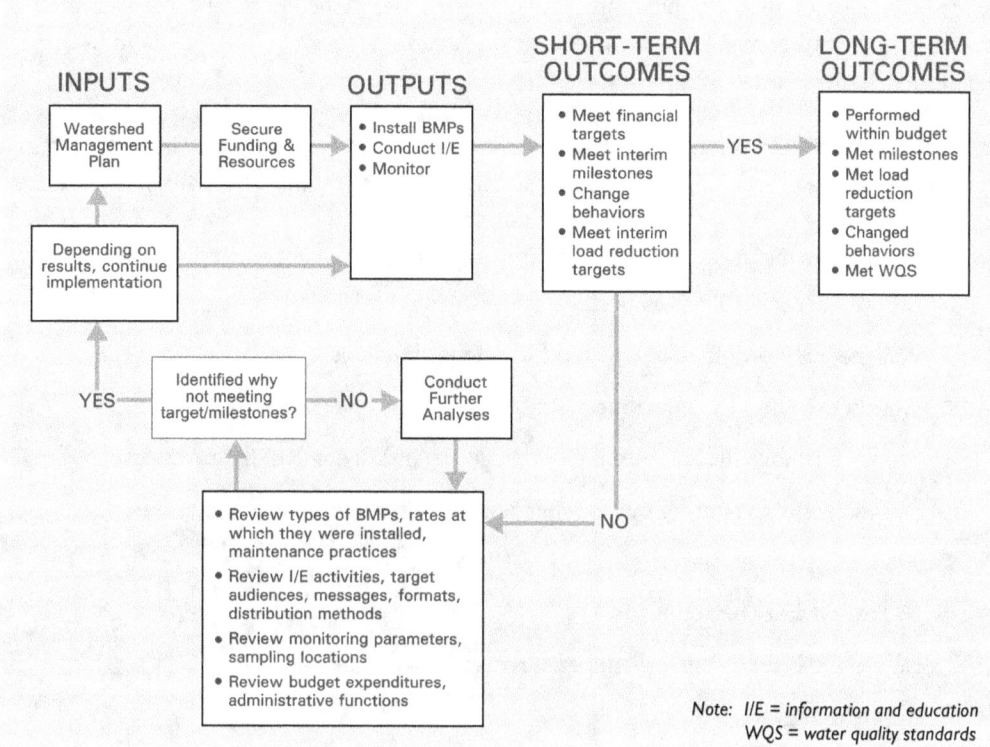

Figure II-6. **The iterative process of the watershed approach.**

As part of developing your implementation plan, you devised a method for tracking progress. Using that tracking system, you should review the implementation activities outlined in your work plan, compare results with your interim milestones, provide feedback to stakeholders, and determine whether you want to make any corrections.

In addition, you will have to analyze your monitoring data. At a minimum you should conduct a routine summary analysis that tracks progress; assesses the quality of data relative to measurement quality objectives (i.e., whether the data are of adequate quality to answer the monitoring question); and provides early feedback on trends, changes, and problems in the watershed. Routine data analysis in this context does not have to be complex or sophisticated. The primary goals are to make sure that your monitoring effort is on track and that you get a general sense of what's going on in the watershed.

Because many watershed activities can affect NPS loads, you should pay attention to broad watershed land use patterns such as overall land use change (e.g., abandonment of agricultural land, timber harvest, large urban development); changes in agriculture, such as acres under cultivation or animal populations; and changes in watershed population, wastewater treatment, stormwater management, and so forth. An annual look at watershed land use is probably enough in most cases.

If it is determined that you are not meeting the implementation milestones or interim targets that you set for load reductions and other goals, what should you do? There are many reasons why you might not see the results you were hoping for, and many of them are easily remedied. Consider the following questions before making any changes to your watershed plan:

- Did weather-related causes postpone implementation?

- Was there a shortfall in anticipated funding for implementing BMPs?

- Was there a shortage of technical assistance?

- Did we misjudge the amount of time needed to install some of the practices?

- Did we fail to account for cultural barriers that prevented implementation?

- Are we implementing and using the BMPs correctly?

- Has the weather been unusual?

- Have there been unusual events or surprises in the watershed?

- Are we doing the right things?

- Are our targets reasonable?

- Are we monitoring the right parameters?

- Do we need to wait longer before we can reasonably expect to see results?

If you have ruled out all the above possibilities, you need to consider whether the plan has called for the right BMPs. It is possible that identifying the causes and sources of pollutants earlier in the planning process was not completely correct or that the situation has changed.

Nine Elements of a Watershed-Based Plan

As earlier mentioned, EPA has identified nine elements considered essential for WBPs. Figure II-7 shows how the nine elements fit into the six steps in the watershed planning and implementation cycle. Why these nine elements? Historically, implementing WBPs with these nine minimum elements has been successful because the plans contain the right balance of scientific input, public outreach, and long-term monitoring. Implementing an effective plan requires tribes and other entities to address all nine elements. The components highlighted in white in the diagram are the nine elements that EPA considers essential in an effective WBP.

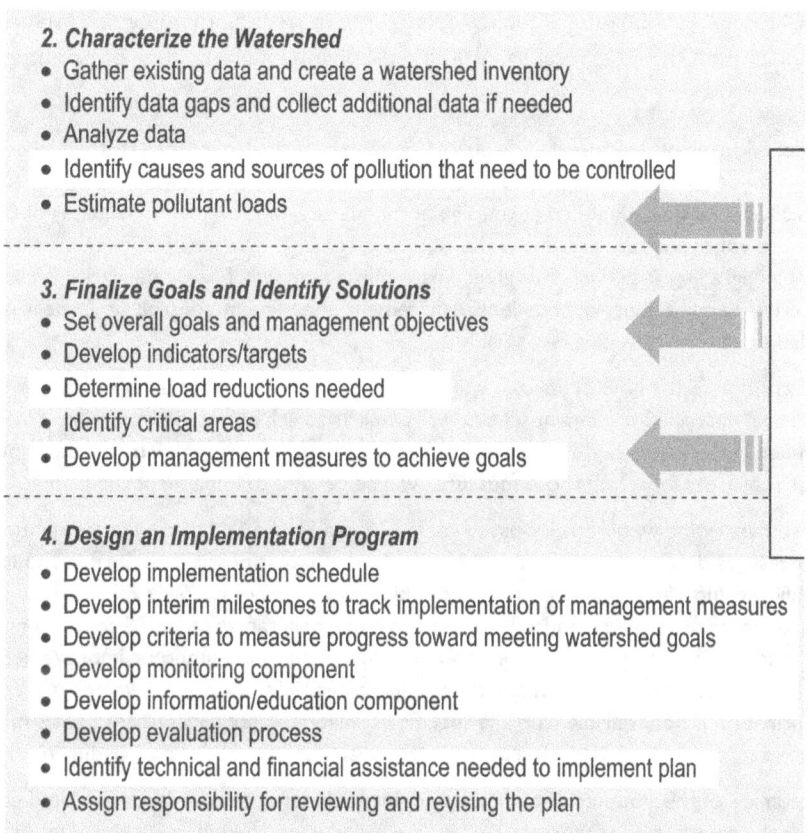

Figure II-7. Source of the nine required elements of a Watershed-Based Plan.

Language related to these nine elements that is specific to tribes is in both the most recent guidelines for tribes and in the annual competitive RFPs. To access these guidelines, go to *www.epa.gov/nps/tribal.*

Note that the requirements established by EPA for tribes and states differ slightly. States are required to estimate pollutant loads and set load reductions. Tribes have the option of estimating pollutant loads and associated reductions or basing their plans on water quality-based goals. EPA has incorporated this flexibility for tribes recognizing that not all tribes have yet developed water quality standards, and many tribes might need additional time or technical assistance to develop more sophisticated estimates of the NPS pollutants that need to be addressed. However, EPA encourages tribes to develop load reductions where that information is available.

For easy reference, the following is the exact language describing the nine essential elements of WBPs, taken from the most recent EPA guidance document at the time of printing.

a. An identification of the causes and sources or groups of similar sources that will need to be controlled to achieve the goal identified in element (c) below. Sources that need to be controlled should be identified at the significant subcategory level with estimates of the extent to which they are present in the watershed (e.g., X number of dairy cattle feedlots needing upgrading, including a rough estimate of the number of cattle per facility; Y acres of row crops needing improved nutrient management or sediment control; or Z linear miles of eroded streambank needing remediation).

b. A description of the NPS BMPs that will need to be implemented to achieve a water quality-based goal described in element (c) below, as well as to achieve other watershed goals identified in the watershed-based plan, and an identification (using a map or a description) of the critical areas for which those measures will be needed to implement the plan.

c. An estimate of the water quality-based goals expected to be achieved by implementing the measures described in element (b) above. To the extent possible, estimates should identify specific water quality-based goals, which may incorporate, for example: load reductions; water quality standards for one or more pollutants/uses; NPS total maximum daily load allocations; measurable, in-stream reductions in a pollutant; or improvements in a parameter that indicates stream health (e.g., increases in fish or macroinvertebrate counts). If information is not available to make specific estimates, water quality-based goals may include narrative descriptions and best professional judgment based on existing information.

d. An estimate of the amounts of technical and financial assistance needed, associated costs, and/or the sources and authorities that will be relied upon to implement the plan. As sources of funding, Tribes should consider other relevant Federal, State, local and private funds that may be available to assist in implementing the plan.

e. An information and education component that will be used to enhance public understanding and encourage early and continued participation in selecting, designing, and implementing the NPS BMPs that will be implemented.

f. A schedule for implementing the NPS BMPs identified in the plan that is reasonably expeditious.

g. A description of interim, measurable milestones for determining whether NPS BMPs or other control actions are being implemented.

h. A set of criteria that can be used to determine whether the water quality-based goals are being achieved over time and substantial progress is being made towards attaining water quality-based goals and, if not, the criteria for determining whether the watershed-based plan needs to be revised.

i. A monitoring component to evaluate the effectiveness of the implementation efforts over time, measured against the criteria established under element (h) above.

Remember, each of the nine elements contained in your watershed plan comes from one of the six steps of watershed planning and implementation discussed above. Some of the elements are stepwise and others you do not need to do in order. The key is to have them all in your WBP. Also, if you are working on NPS programs, many of those elements are things you are already doing. You do not have to reinvent the wheel. Borrow as much as you can from existing documents. For example, you probably are already monitoring, and you already have set milestones, but you might not yet have a formalized education and outreach program. Many of the elements include pieces that you are most likely already undertaking for other programs when you embark on a 319 project. The remainder of this section takes a close look at each of the nine elements and offers examples for the reader.

The goal of this section is to demonstrate that WBPs are worth the effort and that resources exist to help tribes successfully develop and implement them. In addition, EPA encourages tribes to develop partnerships both within the tribe and beyond, in the name of making real, positive changes to their water quality.

Element a. *An identification of the causes and sources or groups of similar sources that will need to be controlled*

For the first of the nine elements, you will need to identify the causes and sources of pollutants or groups of similar sources that you will aim to control to achieve load reductions, and any other water quality goals identified in the watershed plan. As you work through step 1 of the six steps of the watershed planning and implementation process, assembling existing and possibly new data and conducting analyses on your data, you will be working toward this element. Such an analysis will evaluate spatially (e.g., where are the problems?) and temporally (e.g., are the impacts seasonal?). What other trends and patterns, such as land use activities, are occurring?

Perhaps at the beginning of the process, you will be aware of only the broadest categories of NPS pollution that might be responsible for your water impairments—urban runoff, silviculture, construction, or agriculture, for example. At the end of the process, you will able to go beyond the basic categories and state with some certainty that your impairments can be traced to pastureland, rangeland, feedlots, irrigated crop production, forest management, or land development. All these are examples of subcategories of NPS pollution.

But how might you arrive at not only knowing what the pollutants are (the *causes*) but also where the pollutants are from (the *sources*)? The idea is to start with the basics and work your way to specifics. Are you aware of cattle having access to the stream? Is a major construction project going on? Have you seen fish kills in certain tributaries of the waterway? In the initial stages of watershed planning, many of the links might not be thoroughly understood; they will more likely be educated guesses that generate further analyses to determine validity. The key in this step is to work with stakeholders to gather enough information that you can begin to make some educated guesses about where the problems are coming from. Table II-4 lists some of the most common stressors, or pollutants, experienced in watersheds and the likely sources for those stressors. As you would expect, they include some of the most common problems: nutrients, sediment, pathogens, and the like. If you have done an NPS assessment report, this information is already available for reservation waters.

Table II-4. Common pollutant types and sources found in watersheds

Stressors	Sources
Sediment	Row crop land Timber harvest areas Eroding stream banks Construction
Nutrients	Livestock feeding areas Fertilized cropland Septic systems
Bacteria	Livestock feeding areas Septic systems Geese and other wildlife

As you move through the watershed planning process, it is often useful to diagram the links and to present them as a picture or as a conceptual model (see Figure II-8). These diagrams provide a graphic representation that you can present to stakeholders, helping to guide the subsequent planning process. In many cases, there will be more than one pathway of cause and effect. You can also present that concept to stakeholders verbally, as "if…, then…" links. For example, "If the area of impervious surface is increased, flows to streams will increase. If flows to streams increase, the channels will become more unstable." Or, in the simplified conceptual model below, if forest is converted to crop land, more sediment will become available and wash into the river.

The conceptual model can be used to start identifying relationships between the possible causes and sources of impacts seen in the watershed. You do not have to wait until you have collected additional information. In fact, the conceptual model can help identify what types of data you need to collect as part of the characterization process. You will want to make sure that as part of the characterization process you take a trip with the stakeholders for a visual inspection of the waterway. Take pictures and document what you see.

It is critical to go through the process of identifying causes and sources to the best of your ability for a few reasons:

- To increase confidence that costly remedial or restoration efforts are targeted at factors that can truly improve biological condition
- To identify causal relationships that are otherwise not immediately apparent
- To prevent biases or lapses of logic that might not be apparent until a formal method is applied

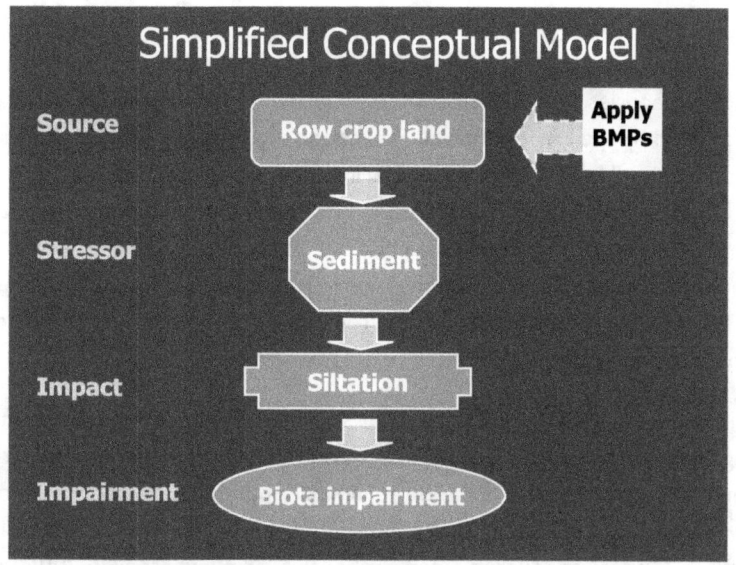

Figure II-8. Simplified conceptual model.

Figure II-9 demonstrates how this deductive process fits into the overall planning and implementation program and illustrates the importance of this step to the overall result.

For a detailed description of the stressor identification process, see EPA's *Stressor Identification Guidance Document* (*www.epa.gov/waterscience/biocriteria/stressors/stressorid.html*). In addition, two stressor identification modules originally developed as part of EPA's 2003 National Biocriteria Workshop are available online. The SI 101 course contains several presentations on the principles of the stressor identification process: *www.epa.gov/waterscience/biocriteria/modules/#si101*.

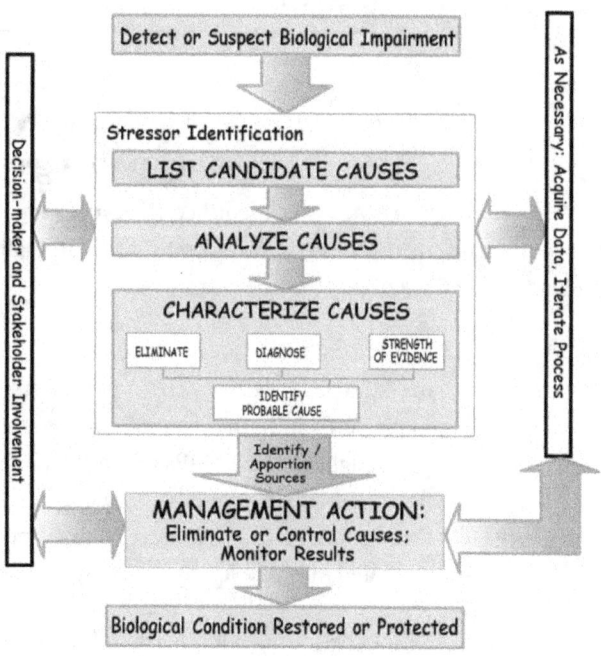

Figure II-9. Role of deductive process in program planning and implementation.

Example: Identification of Causes and Sources (JST 2007):

Causes and sources of pollution in the Dungeness Watershed area include most of the eight major categories of nonpoint source pollution in the waterways of the United States identified by the Environmental Protection Agency: Agriculture; Forestry; Hydromodification/Habitat Alteration; Marinas and Boating; Roads, Highways and Bridges; Urbanization; and Wetland/Riparian Management; with subcategories specific to each. Of the EPA's major categories, all but one (mining) are relevant to the Dungeness Watershed. The history of collaborative watershed planning in the Dungeness region has resulted in an extensive collection of data regarding the sources and trends of pollution in each of these categories.

The Agriculture category in EPA's nonpoint source classification system contains three sub-categories relevant to the Dungeness watershed: irrigated crop production, animal operations, and aquaculture. Of the three, animal operations present the most significant risk to water quality and human health in the Dungeness Watershed. The Clallam Conservation District conducted a comprehensive farm inventory in 2006 utilizing air photos and visual inspection of farms from the road. Of the 125 farms in Clallam County that were rated as having a medium or high potential to impact surface water quality, 23 were located in the Dungeness drainage covering 495 acres.

Element b. Description of NPS BMPs to achieve water quality-based goals

Once you have identified the causes and stressors, you can select the optimal ways to address the problems. BMPs can be structural, such as silt fences, or nonstructural, such as changing grazing patterns.

It is critical that your BMPs correspond to the sources of your impairments and that they are measurable. Specify where in the watershed they will be placed, and carefully consider access to property and other factors when you determine the locale of your BMPs. For example, your plan might say that BMPs will be implemented "on all abandoned mine sites with dry-weather flows" or "on all stream banks along upper reaches" or "on all livestock facilities on Willow Run." A map will be very useful in demonstrating where BMPs will be used or installed and relocated, if possible.

What does it mean to say that BMPs should be linked to (or otherwise address) stressors and sources? It means that *water quality goals or estimates for pollutant removal rates should be included in your BMP design*. If you are not estimating load reductions, these can be based on typical ranges, i.e., percentage removed or treated and other reasonable estimates.

So how do you select the best BMPs for your site? Select BMPs that make sense for solving the problem you have identified at your location. There are numerous resources to draw on in determining which BMPs will achieve the best results for the site, stressors present, and sources identified. The cover of one popular source, EPA's *National Management Measures for the Control of Nonpoint Pollution from Agriculture* is featured to the right. On the Web site *www.epa.gov/nps/pubs.html* there are also manuals for hydromodification control BMPs, urban area BMPs, BMPs for restoring wetlands, and much more (see direct links in Table II-5). Nearly all the resources you need are on the Internet. Both EPA and USDA have comprehensive BMP manuals that are available online. Another excellent source of BMP information, including photos, are at *www.waterquality.utah.gov/TMDL/Virgin_River_Watershed_Implementation_Appendix.pdf*. Many other states also have information on BMPs.

Table II-5. Online resources for BMPs

BMP technical area	Online resource
Agriculture	*www.epa.gov/nps/agmm*
Forestry	*www.epa.gov/nps/forestrymgmt*
Hydromodification	*www.epa.gov/nps/hydromod*
Marinas and recreational boating	*www.epa.gov/nps/mmsp*
Urban areas	*www.epa.gov/nps/urbanmm*
Wetlands and riparian restoration	*www.epa.gov/nps/wetmeasures*
Coastal waters	*www.epa.gov/nps/MMGI*

It is often recommended that you evaluate which BMPs to use by ranking them. Take a look at the BMPs that exist for your subcategory of sources of NPS pollution, and then narrow the list down from many to a few until you find the ones that have the right combination of effectiveness and ease of utility for your particular site. Consider factors like those listed below when selecting BMPs:

- Importance of waterbody: Is it a drinking water source or a cultural or recreational resource?

- Magnitude of impairment(s)

- Level of effort needed

◻ Existing loads (causes and sources)

◻ Magnitude, spatial variation, clustering

◻ Ability of BMPs to reduce loads

◻ Certainty of success of your BMP

◻ Feasibility of implementation: Do you have willing partners? Public support? Is the landowner willing to do this?

◻ Additional benefits, recreational enhancements, demonstration

Another tool that can be useful for selecting your BMPs is a BMP Effectiveness Table. Shown below (Table II-6) is one of the BMP effectiveness tables in the agricultural BMPs document described above and posted on EPA's Web site. It lists various agricultural management practices in the left column. The other columns provide some relative effectiveness information for each BMP (as an individual unit) for controlling runoff volume, phosphorus, nitrogen, sediment, and bacteria. Tables like that, which aggregate BMP performance, can be used to screen various management practices and to provide information on what might work in a particular situation to reduce effects of agricultural operations. Remember that they are estimates per unit and that actual effectiveness on the ground will be determined by the effectiveness of placement and operation of all the BMP units collectively. Nevertheless, tables like this one offer examples of the effectiveness of certain BMPs (called Practice Category in the example below) on reducing total phosphorus, total nitrogen, sediment, and fecal coliform. Information like that can help you visualize which BMP might be the most effective; however, you must still weigh other factors and determine what is right for your program.

Table II-6. Example of a BMP effectiveness table

Practice[b] Category	Runoff Volume	Total[d] Phosphorus (%)	Total[d] Nitrogen (%)	Sediment (%)	Fecal Coliform (%)
Animal Waste Systems[e]	reduced	90	80	60	85
Diversion Systems[f]	reduced	70	45	NA	NA
Filter Strips[g]	reduced	85	NA	60	55
Terrace System	reduced	85	55	80	NA
Containment Structures[h]	reduced	60	65	70	90

NA = not available.
[a] Actual effectiveness depends on site-specific conditions. Values are not cumulative between practice categories.
[b] Each category includes several specific types of practices.
[d] Total phosphorus includes total and dissolved phosphorus; total nitrogen includes organic-N, ammonia-N, and nitrate-N.
[e] Includes methods for collecting, storing, and disposing of runoff and process-generated wastewater.
[f] Specific practices include diversion of uncontaminated water from confinement facilities.
[g] Includes all practices that reduce contaminant losses using vegetative control measures.
[h] Includes such practices as waste storage ponds, waste storage structures, waste treatment lagoons.

Source: www.epa.gov/nps/agmm

A key point to keep in mind: See what groups are already doing in your watershed. Maybe they are doing something that is working well already. Once you have reviewed ongoing activities, take a look at those online references. You will want to be familiar with what activities are already underway before embarking on resource-intensive projects. Also remember to always consider:

☐ What is essential to achieving objectives?

☐ Which options do the stakeholders prefer?

☐ Which options have greatest chance for long-term success and sustainability?

Not every BMP will be right for everybody. Consider the unique factors in your watershed and choose accordingly.

Example: BMP Table (JST 2007)

The following table illustrates the activities that the Jamestown S'Klallam Tribe will undertake to implement its BMPs:

Table II-7. Summary of NPS management measures in the strategy and implementation plan *(adapted from Streeter and Hempleman 2004)*

General strategies	Identified actions
Strategies to Address Human Waste	
Expansion of septics operation and maintenance programs	Assessment and monitoring
	Inspect septics of concern
	Followup repair, replacement
Purchase of land and conservation easements in sensitive areas.	River's End area targeted for restoration due to septic failures and critical salmon habitat.
	Land/easement purchase, building demolition, septic removal.
Conversion to community systems where appropriate	Identified areas are 3 Crabs Road, Golden Sands Development, Carlsborg
	Feasibility, design, implementation
Landowner Education	Septics 101 class on basic septic system maintenance.
	School water quality curriculum
	Talks and displays at River Center, displays at area festivals and events
Stormwater Management	
Sub-area plans for stormwater management	Focus areas are Marine Drive, 3 Crabs Area
	Restoration of hydrological function in Meadowbrook Creek
	Capital facilities, retrofits, standards for new development, and basic BMPs based on soil characteristics, topography, and development patterns.

Table II-7. Summary of NPS management measures in the strategy and implementation plan *(adapted from Streeter and Hempleman 2004) (continued)*

General strategies	Identified actions
Low Impact Development	CCD stormwater management manual for small-scale development (rural residential) in progress. The manual includes a series of pre-engineered stormwater management practices for builders, developers and citizens which can be installed without the aid of an engineer.
	The North Peninsula Builders Association has developed a Built Green Checklist
	Landowner education.
Agriculture and Livestock Waste	
Treatment of irrigation ditch tailwaters	Pilot projects completed, biofiltration, constructed wetlands. Marine Drive specifically identified for treatment.
Ditch piping	Reduction of bacterial contamination through piping of open ditches, based on priorities identified in CCD monitoring
Individual conservation plans and BMPs	CCD activities based on 2006 farm inventory.
Outreach and education	Workshops and presentations
	Brochures such as "Living on a Ditch"
	Web page information
Enforcement	WA Dept of Ecology per MOU with Clallam County and Clallam Conservation District
Domestic Animals and Pet Waste	
Public outreach	Waste disposal information via brochures, advertisements and presentations; signage
Installation of pet waste stations	Areas of high pet use adjacent to surface waters
Cleanup	Coordination of volunteer cleanup crews
Regulatory and Policy Approaches	
Stormwater ordinance or designation of stormwater sensitive areas	Revisions to draft; proceed to adoption (Clallam County)
Critical Areas	Update maps and regulatory constraints per Federal ESA listings and WA Legislature action
Review development regulations	Encourage use of LID, remove disincentives.
Establish Tribal regulations	Adopt ordinances to regulate activities on Tribal reservation/trust lands
Research and Monitoring	
Freshwater	Develop overall freshwater monitoring for wet season/storm events for streams, ditches
	Continued fresh water monitoring
	BMP effectiveness monitoring
	Data analysis of monitoring
	Microbial source identification
	Streamkeeper voluntary monitoring program

Table II-7. Summary of NPS management measures in the strategy and implementation plan *(adapted from Streeter and Hempleman 2004) (continued)*

General strategies	Identified actions
Marine and estuarine areas	Continued marine monitoring
	Beach and Beachwatcher voluntary monitoring programs
	Microbial source identification
	Additional research on Dungeness Bay (basic ecological studies, nutrients, circulation, fecal coliform assessment in water and sediment, wildlife usage)
	Feasibility for remediation of Meadowbrook/Cooper/Matriotti/Gierin creek estuaries
	Marine shoreline soft armoring techniques
General	Analysis of impervious surfaces
	GIS analysis, map fecal nutrient and temporal trends
Education and Outreach	
General outreach	Public workshops
	Newspaper reports
	Continuation of school age water quality classes and field trips
	Displays and activities in booths, fairs and festivals
	Permanent displays at River Center
	Web pages at Tribe and River Center

Element c. An estimate of water quality-based goals expected to be achieved

Now that you have selected what you consider to be the best BMPs to address your issues in an acceptable, feasible, and efficient manner, what kinds of results do you expect to see?

States typically think of load reductions in terms of setting goals. A load estimate tells you how much of each pollutant, such as sediment, phosphorus, or nitrogen, is being loaded into the waterbody by each group of sources, such as row crops or forest roads. Usually this is done using computer modeling programs. That information is used to determine by how much each pollutant type entering the waterbody must be reduced.

Tribes are not required to comply with the load estimation aspects of the nine elements. Therefore, for tribes that do not choose to go the route of estimating or quantifying pollutant loads, the process is somewhat different. After identifying causes and sources of the stressors, instead of estimating loads, tribes are required to produce an estimate of their water quality-based goals (quantitatively or narratively). These can be defined as the following:

- Load reductions
- Water quality standards for one or more pollutants/uses
- NPS TMDL allocations
- Measurable, in-stream reductions in a pollutant
- Improvements in a parameter that indicates stream health (e.g., increases in fish or macroinvertebrate counts, habitat improvements)

If information is not available to make specific estimates, water quality-based goals may include narrative descriptions and best professional judgment based on existing information.

When you are setting goals and identifying solutions, you will need to tell EPA what you want to achieve, where it is that you want to achieve it (your critical areas), and how you want to do so (your BMPs).

To summarize, in the last section you identified stressors and sources to be controlled and estimated how much of your pollutant of concern was entering a selected waterbody. Once you have identified your particular stressors, you can check what the water quality-based goals for those pollutants are. These can be general goals, such as reducing the amount of sediments in a portion of the waterway, or specific load reductions if available. The next step is to match the right BMPs to mitigate the problems and to prioritize your critical areas.

Example from Karuk Tribe, Department of Natural Resources Eco-Cultural Resources Management Plan (KT 2006)

The Karuk Tribe desires the implementation of methods to limit and/or mitigate for the sediment transport or delivery of materials which degrade water quality and fisheries habitat. Where feasible, areas contaminated with mercury or other toxins should be located, decontaminated, and restored. Additionally, inactive mines should be properly contained to prevent off-site transport of material or contamination of ground and surface waters. Limit the used of suction dredging in rivers and creeks at times that threaten fisheries or water quality.

There is also a need to restore hydraulic mine areas in many instances, these areas are directly adjacent to watercourses. These areas do not maintain a significant vegetation component and subsequently can contribute to excess heating of adjacent streams.

Objectives:

Implement restoration measures that mitigate damaged areas affected by past hydrologic mining to minimize soil erosion, reconfigure topographic contours and drainage, and manage vegetation to enhance the structure and composition to accommodate natural processes (fire, hydrologic connectivity, and nutrient cycling). Remove and/or reduce the presence of toxins such as mercury, sulfuric acid and cyanide in sediment deposits and watercourses. Monitor and reduce the effects and activities associated with suction dredge mining along the Klamath and Salmon River watersheds. Inventory rock sources and mitigate for erosion potential and off site sediment delivery. Develop economically and environmentally low impact methods of aggregate removal to supply for local upgrade, maintenance and restoration activities. Work with Federal, State, and County Agencies, and community groups to ensure cultural/natural resource protection measures are adequate and in place.

Element d. *An estimate of the amounts of technical and financial assistance needed, associated costs needed to implement your plan*

A critical factor in turning your watershed plan into action is the ability to fund implementation. Funding is needed for all the activities in the WBP, such as management practice installation, information and education activities, monitoring, and administrative support. In addition, you should document the types of technical assistance needed to implement the plan and the resources or authorities that will be relied on for implementation, in terms of both initial adoption and long-term operation and maintenance (O&M). For example, if you have identified adoption of tribal ordinances as a management tool to meet your water quality goals, you should involve the local authorities that are responsible for developing those ordinances and outline their short-term and long-term roles.

The estimate of financial and technical assistance should take into account the following:

- Administration and management services, including salaries, regulatory fees, and supplies, as well as in-kind services efforts such as the work of volunteers and the donating of facility use

- Information/education efforts

- The installation, operation, and maintenance of BMPs

- Monitoring, data analysis, and data management activities

Haaku Water Office restoration project near Sky City, Acoma Pueblo.

Some of the costs of implementing your watershed plan can be defrayed by leveraging existing efforts and seeking in-kind services. Some examples follow.

Use existing data sources. Most geographic areas have some associated background spatial data in the public domain, such as digital elevation models, stream coverages, water quality monitoring data, and land cover data. Note that the EPA Quality System (EPAQA/G-5; *www.epa.gov/quality*) recommends that a QAPP be prepared for using existing data and for collecting new data.

Use existing studies. Many agencies have reports of previous analyses, providing useful baseline information and data, such as delineated subwatersheds or a historical stream monitoring record. The analyses might have been done for another purpose, such as a study on fish health in a stream, but they can contribute to an understanding of the background of the current concerns.

Use partnerships. State, county, or federal agencies working as technical assistance providers and implementing natural resource program initiatives can offer computer services and expertise, such as performing GIS analysis or weaving together elements of different programs that might apply to the local area. They might be in a position to write part of the overall watershed plan if they have existing generalized watershed characterization studies.

Cover incidental/miscellaneous costs through contributions. For example, staff time to assemble needed elements, supplies, and meeting rooms for a stakeholder or scoping meeting can all be donated. As a start, refer back to the checklist you compiled from your stakeholder group earlier to determine what resources are available in the group.

Locating Federal Funding

For a complete list of federal funding sources, visit the *Catalog of Federal Domestic Assistance* (*www.cfda.gov*). That Web site provides access to a database of all federal programs available.

Also visit *www.epa.gov/watershedfunding* to view the *Catalog of Federal Funding Sources for Watershed Protection*. That interactive Web site helps match watershed project needs with funding sources.

Amber Marriott

Example: Excerpt from the Dungeness WBP Estimated Budget (JST 2007)

Table II-8. Overview of estimated costs

Notes on this table: Cost estimates are organized by Tribal sub-goal and were obtained from a variety of sources including the Tribal NPS Management Plan, Clean Water Strategy/DIP, Comprehensive Irrigation Water Conservation Plan, Salmon Recovery Plan, and partner organizations such as Clallam Conservation District and Dungeness River Audubon Center. Some estimates represent ongoing program costs, while others are one-time project costs, thus they have not been added together to present a total annual or 5-year estimate. Where appropriate, the estimates have been separated for Tribal and partner NPS programs.

Activities by tribal goal	Estimated costs	Time frame	Comments	Funding sources
WATER QUALITY MANAGEMENT MEASURES				
Tribal NPS Programs				
Staff and office supplies to oversee monitoring of water and shellfish, septic programs, stormwater planning, revegetation and other BMPs for protecting and improving water quality. 4.75 FTE	$522,000	annual costs, ongoing	details in Tribal NPS plan; includes staff, office supplies, travel, training, overhead, and engineering/ tech consultation	EPA, BIA/ OSG, Ecology, WDFW, Tribal Funds
Projects for channel restoration, stormwater facilities, revegetation, agriculture BMPs, septic remediation	$325,000	annual costs, 5 year plan	project construction costs	
Public education programs	$30,000	annual, ongoing		EPA, Ecology, Foundations
Monitoring supplies, transportation, laboratory costs	$30,000	annual, ongoing		EPA, BIA-OSG, Ecology, WDFW, Tribal Funds
Land acquisition where other water quality remediation is not feasible (acquisition and removal of structures and septic systems)	$250,000	2-10 years		Land and Water Conservation Fund, RCFB, Private
Partner NPS Programs				
Clean Water Strategy				
Ongoing actions for septic inspection and education programs, public outreach, stormwater planning, livestock plans, monitoring, streamkeepers	$2,900,000	5 years	Clean Water Strategy/DIP has detailed breakout	Centennial Fund, EPA, Special Assessments, County Fees, WRIA 18 implementation funds, other

Element e. An information and education component that will be used to enhance public understanding

An important aspect of pollution prevention is to think ahead and make people aware of their impacts on water quality. When people are made aware of the consequences of their actions, they might start to think about what they can do to mitigate their impacts. The best way to go about raising awareness is to implement a public outreach, education, and involvement plan. The information discussed in this section is from the River Network's Training for Watershed Trainers, Communications Planning Session, December 2007 (*www.rivernetwork.org*).

An outreach, education, and involvement plan will help you and your organization educate people on water quality concerns in your area and their effects on declining water quality. Many people might not realize that their actions can affect water quality. For example, allowing cattle to graze along a nearby spring can cause *E. coli* contamination, and driving across a stream to collect plants for medicinal purposes can cause sedimentation in the stream.

An outreach and education plan can also be used to give the public a heads-up about upcoming NPS projects that they will be seeing throughout the reservation or watershed. If a passer-by sees a fencing project being implemented at a popular fishing spot, he is more likely to respond positively to the project if he already knows about it.

Before developing an outreach plan, ask yourself the following questions:

- What do we have? For example, polluted runoff in waterways from unrestricted livestock grazing.
- What do we want to change? For example, less runoff equals cleaner water quality.
- What is on our wish list? For example, raise awareness to change actions and practices.
- What is our budget? How much money do we have? (Be realistic.)
- What are our communication goals? For example, use mailings, attend public meetings, and have a booth at the Environmental Department Annual Earth Day Fair.

Audience Development

An important part of your audience is the stakeholders. A stakeholder is a group or individual who has the responsibility for implementing the decision, is affected by the decision, or has the ability to impede or assist in implementing the decision. Stakeholders are an important part of the audience because you want to ensure that their concerns are factored into the decisions made to address water quality pollution. In addition, they have the ability to impede the effectiveness of any projects carried out if their concerns are not considered.

Once you have determined your outreach plan, focus on a segment of your targeted audience. Ask yourself these questions:

- Who needs to hear your message?

- Who has influence over your targets?

Then get to know your audience by listening to them. Invite a select group of people fro your targeted audience, such as elders or tribal council members, to a meeting or a dinn Also, use this opportunity to ask if they have their own priorities and issues and see if you can meet somewhere in the middle.

Message Development

Next, develop your message using feedback from your stakeholder group and use one of four ingredients of a good message: value, barrier, ask, or vision.

Value   For example, a tribal elder reminisces about how long ago as a little boy he used to swim at a nearby swimming hole with his grandfather, but now he cannot swim with his grandchildren because the water is so dirty. The poor water quality is from cattle overgrazing at the swimming hole and lack of vegetation and shade around the swimming hole. The water quality has declined because of high water temperature from lack of shade and high levels of fecal coliform bacteria from cattle. The value message helps the audience to relate and care about the issue because it touches on a family value.

Barrier   For example, a tribal elder reminisces about how long ago as a little boy he used to swim at a nearby swimming hole with his grandfather, but now he cannot swim with his grandchildren because cattle waste has rendered it too dirty. The barrier message helps the audience relate and realize that something can be done about it.

Ask   For example, a tribal elder reminisces about how long ago as a little boy he used to swim at a nearby swimming hole with his grandfather, but now he cannot swim with his grandchildren because cattle waste has contaminated the water. The tribal elder asks the audience, "Wouldn't you want your grandchildren to stop playing video games at home and go out and spend some time swimming with you?" The ask message helps the audience to start thinking about what can be done to fix this problem and what the positive effects could be.

Vision   For example, a tribal elder reminisces about how long ago as a little boy he used to swim at a nearby swimming hole with his grandfather, but now he cannot swim there with his grandchildren because the cattle that use the swimming hole for drinking leave the water too dirty for swimming. He imagines how it would be fun to teach his grandchildren how to swim and to be there when they start to swim on their own. The vision message helps the audience clearly envision what it would be like to have clean water again.

When you develop a message to use in your outreach activities, remember to use words your targeted audience will understand. Avoid using technical terms. For example, not many people know what *nonpoint source pollution* means, so use *polluted runoff* in your message instead.

Delivering Your Message

After you have developed your message, you will want to practice it. Meet with a couple people from your stakeholder group and ask them to listen for the following when you practice your message:

- Content: Is it relevant to them?

- Words used: Are they able to understand?

- Engagement: Are you captivating their interest?

- Impact or pressure: Is the amount just right or too much pressure?

- Organization: Does your message make sense?

- Timing: Is it too long or too short?

Now that you have developed your message and know your audience, you are ready to deliver your message. Many methods to deliver your message are available. Develop brochures, fact sheets, and door-to-door materials to be used for in-person outreach such as fairs, events, hands-on demonstration projects, school presentations, tribal council presentations, and public tours of project sites. Media resources—Web site postings, newsletters, articles in the newspaper, and radio spots—can also be used as methods of delivering your message. If one method does not work, use another until you find something that works.

Where to Get More Help on Information/Education Activities

For more information on planning and implementing outreach campaigns, refer to EPA's *Getting in Step: A Guide for Conducting Watershed Outreach Campaigns.* This comprehensive guide will walk you through the six critical steps of outreach—defining your goals and objectives, identifying your target audience, developing appropriate messages, selecting materials and activities, distributing the messages, and conducting evaluation at each step of the way. You can download the guide at *www.epa.gov/owow/watershed/outreach/documents/getnstep.pdf* or order it by calling 1-800-490-9198.
Ask for publication number EPA 841-B-03-002.

Don't Reinvent the Wheel

EPA has developed a *Nonpoint Source Outreach Digital Toolbox*, which provides information, tools, and a catalog of more than 700 outreach materials that state, tribal, and local agencies and organizations can use to launch their own NPS pollution outreach campaigns. The toolbox focuses on six NPS categories: stormwater, household hazardous waste, septic systems, lawn care, pet care, and automotive care, with messages geared to urban and rural settings. Outreach products include mass-media materials, such as print ads, radio and television public service announcements, and a variety of materials for billboards, signage, kiosks,

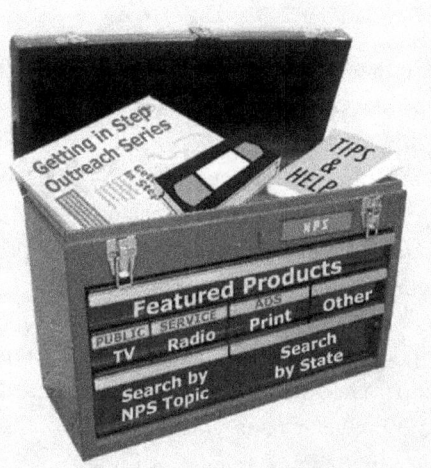

posters, movie theater slides, brochures, factsheets, and everyday object giveaways that help to raise awareness and promote non-polluting behaviors. The outreach products can be easily modified for use in raising awareness about other NPS categories not discussed here. The toolbox is available online and as a CD at *www.epa.gov/nps/toolbox*.

Example: Karuk Tribe's Plan (KT 2006)

Environmental Education has been very important to the Karuk Tribe since program inception. Environmental Education projects serve to inform Tribal and local community members about the Department's mission. Projects such as Fall Salmon Spawning Surveys, during which students collect data that is used by the California Department of Fish and Game, not only give these students hands-on training, but encourage a deeper appreciation of natural resources and ecological processes. The Department's Environmental Education Program provides opportunities for people to correlate current science with traditional knowledge and cultural practices.

Objectives:

Instill in students and adults a life-long desire to learn about and care for their environment. Provide opportunities for youth to learn from Tribal elders about traditional Karuk land and resource management practices. Work with local schools, agencies, organizations, community groups and Tribal members to enrich student and adult knowledge of local environmental and watershed issues to ensure protection of cultural/natural resources. Implement and assist with projects on recycling, community gardening, salmonid spawning and habitat needs, ethnobotany, and other relevant environmental issues to teach students to be good stewards of their local resources and ecological processes. Train students and adults to put their knowledge into practice by providing hands-on activities both in classrooms and outdoors.

Element f. A schedule for implementing the NPS BMPs identified in the plan that is reasonably expeditious

The schedule component of a watershed plan involves turning goals and objectives into specific tasks. The schedule should include a timeline showing when each phase of the step will be implemented and accomplished, as well as the agency/organization responsible for implementing the activity. In addition, the schedule should be broken down into increments that you can reasonably track and review. For example, the time frame for implementing tasks can be divided into quarters. You will prepare more detailed schedules as part of your annual work plans.

In developing schedules, it helps to obtain the input of those who have had previous experience in applying the recommended actions. Locate experienced resource agency staff and previous management practice project managers where possible to identify the key steps. Be sure to note sequence or timing issues that need to be coordinated to keep tasks on track. For example, your project might require applying for permits, which should be accounted for in the implementation schedule. As part of your implementation program, you should set some criteria by which to determine whether you are achieving load reductions or water quality goals over time and making progress toward meeting your overall watershed goals. These criteria can also support an adaptive management approach by providing mechanisms by which to reevaluate implementation plans if you are not making substantial progress toward meeting your watershed goals.

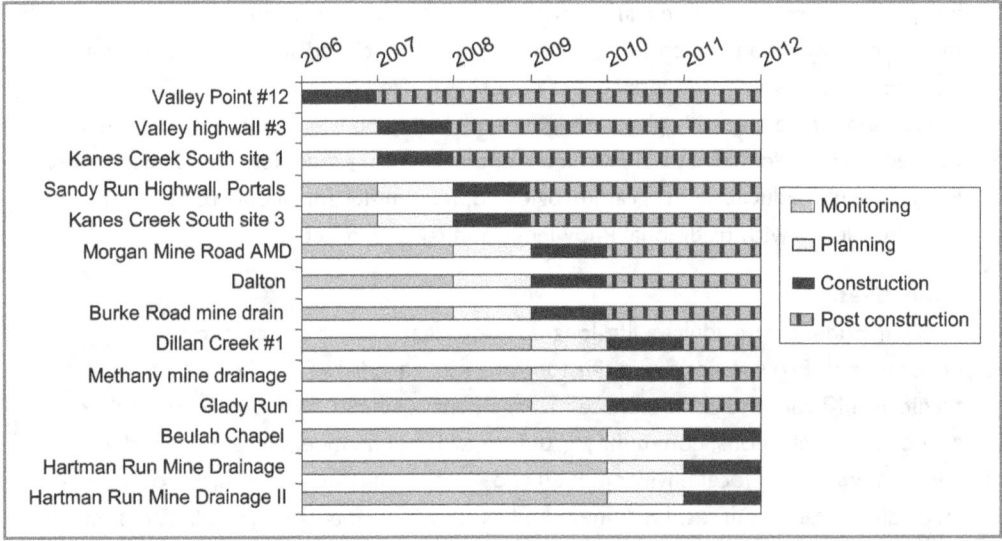

Figure II-10. Implementation schedule for high-priority AMD sources.
Source: Watershed Based Plan for the Deckers Creek Watershed, Preston and Monongalia Counties, West Virginia, August 2006.

Element g. *A description of interim, measurable milestones for determining whether NPS BMPs or other control actions are being implemented*

One means of supporting detailed scheduling and task tracking is to identify interim, measurable milestones for determining whether management practices or other control actions are being implemented. What do you want to accomplish by when? It usually helps to think of milestones in terms of relevant time scales. For example:

- Short-term (1 to 2 years)
- Mid-term (2 to 5 years)
- Long-term (5 to 10 years or longer)

It is also helpful to think of the milestones as subtasks, or what needs to be accomplished over time to fully implement the practice or BMP. When determining time scales and subtasks for actions, milestones should be placed in the context of the implementation strategy. Given the selected practices and the available funds or time frame for obtaining grants, estimate what can be accomplished by when.

First, outline the subtasks involved and the level of effort associated with each to establish a baseline for time estimates. Next, identify the responsible parties associated with the steps so that you can collectively discuss milestones and identify those that are feasible and supported by the people who will do the work.

Watershed restoration activities: creating a rock check dam with two rain gardens up and downstream of a dam, with planted sedges (EPA Region 5 NPS workshop, 2007).

Example: Dungeness WBP (JST 2007)

Table II-9 contains an excerpt from the Implementation schedule of the Dungeness Watershed Based Plan (JST 2007).

Table II-9. Milestones for implementation and measurable criteria for evaluating progress

Note: This table has been organized around Jamestown S'Klallam Tribal natural resources sub-goals, shown in the shaded areas, and the NPS categories addressed. Under each goal, the table summarizes management strategies, milestones for implementation, anticipated timing (subject to funding), key watershed partners, how our outputs will be measured, and the criteria to be used to evaluate progress. There is a considerable amount of overlap between tribal goals, and the management strategies might address more than one source of NPS pollution. Thus, some of the major milestones and evaluation criteria might apply to more than one goal and NPS category.

TRIBAL SUB-GOAL: *Ensure water quality that protects fish and wildlife resources and provides safe food and water.*
NPS Categories Addressed: Agriculture, Hydromodification & Habitat Alteration, Urbanization, Marinas & Vessels

NPS management strategy	Milestones for implementation	Timing	Key partners	Measureable outputs	Measureable criteria for evaluating progress (outcomes)
Human Waste Management	Support county programs for septic O&M, septic inspection and remediation	ongoing	County	# septic systems pumped	Reductions in fecal coliform loading to Matriotti Creek, Dungeness River & Bay: achieve 80% of required reduction by 2010, meet water quality standards by 2012
				Database tracking of septic O&M	
				# systems inspected and repaired	
				# classes Septics 101	
	Suppport the development of community systems for areas of concern	5 years	County	Feasibility studies completed for 3 systems, and at least one to design & construction phase.	
	Buyout remaining parcels at Rivers End and decommission septic systems	3 years	Tribe, County, WDFW, NOLT	Parcels purchased and septics decommissioned.	
Stormwater Management	Treatment of irrigation ditch tailwaters	5 years	CCD, WUA	Installed treatment sites	Achieve net reductions in nitrates for Carlsborg and other elevated areas; achieve net reductions in bacteria; achieve public health standards. No increase in metals or hydrocarbons for baseline streams.
	Piping of irrigation ditches	20 years	WUA	# feet ditch lined	
	Increase use of LID methods	5 years	Tribe, County, CCD	County approval of LID techniques	
	Reduce stormwater impacts	1.5 years	County, Tribe	County approval of upgraded stormwater manual; County Roads Dept install LID in roadside ditches	
		5 years	Tribe	Retrofit tribal facilities	
	Raise public awareness	5 years	Volunteers	Signage program at stormwater drains	

Table II-9. Milestones for implementation and measurable criteria for evaluating progress *(continued)*

NPS management strategy	Milestones for implementation	Timing	Key partners	Measureable outputs	Measureable criteria for evaluating progress (outcomes)
Animal Waste Management	Implement pet waste program	1 year	Tribe, EPA, volunteers	Mailings and posters distributed	Reductions in fecal coliform loading per TMDLs
		2 years	County parks	Install pet waste stations	
	Reduce domestic animal waste	5 years	CCD	Complete farm plans with BMPs	
	Enforce animal waste violations	ongoing	Ecology	Last resort after outreach and technical assistance.	
Monitoring	Shellfish sampling	ongoing	Tribe, DOH	Bi-weekly for intertidal harvest during PSP season; weekly for geoduck harvest	Safe consumption of shellfish by tribal citizens and general public
	Marine waters	ongoing	Tribe, DOH	Monthly sampling	Achievement of certified shellfish beds in Dungeness Bay by 2012
	Freshwater	ongoing	Stream-keepers, Bay-watchers, Tribe, County	Monthly sampling; quarterly for some parameters	Safe wading/swimming for tribal citizens and general public; attain temperature/DO targets for fish bearing streams
	Complete annual data analysis	annual	Clean Water Work Group	Annual review of results	Adaptive mgt of strategies based on results.
	Achieve capacity to monitor all nine parameters required under CWA Section 106		Tribe	Trained staff, all equipment available, funding for staff, materials, transportation, lab costs	Attain water quality standards for all nine parameters
Research	Microbial source identification	1 year	Tribe, Battelle, Ecology, County	Identification of controllable sources of bacterial contamination.	Adaptive mgt of strategies based on results.
	Evaluate D Bay for nutrients and wildlife contribution	5 years	Tribe, Ecology, DOE, WDFW	To be developed	
	Evaluate culture methods for oysters, clams etc	ongoing	Tribe, volunteers	Shellfish gardens completed	
	GIS analysis and remote sensing	ongoing	Tribe, CCD, County	Annual airphotos, updated maps	
	Evaluate effectiveness of BMPs	ongoing	Tribe, County, CCD	Progress reports	Adjustment of BMPs to achieve water quality standards

Table II-9. Milestones for implementation and measurable criteria for evaluating progress *(continued)*

NPS management strategy	Milestones for implementation	Timing	Key partners	Measureable outputs	Measureable criteria for evaluating progress (outcomes)
Research *(continued)*	Investigate restoration of pocket estuaries at Meadowbrook, Cooper, Casselary, Gierin Creeks	5 years	CCD, Tribe, WDFW	Complete feasibility analysis and identify restoration options	
Regulatory	Upgrade city and county ordinances	1 to 5 years	County, Ecology	Adopt stormwater manual	Improved water quality(bacteria, nutrients, chemicals)
			MRC, City	Designate nearshore critical areas	Revisions to CAO
			County, Tribe, CCD	Identify barriers to improved water quality in ordinances	Updated ordinances leading to improved water quality
	Develop/update Tribal ordinances	1 to 5 years	Tribe	Improve jurisdictional control over tribal waters	Updated/new ordinances leading to improved water quality
Education and Outreach	Public workshops on water quality issues	annual	County, CCD, River Center, Tribe	# Workshops conducted; # individuals attending	Behavior change leading to improved water quality; participant feedback
	Prepare written material for public outreach	annual	County, CCD, Tribe	Newspaper articles and mailings; annual milestones report of DRMT; # publications distributed	
	Booths, fairs and festivals	biannual	All partners	Dungeness River Festival; attendance and participation	Behavior change leading to improved water quality; participant feedback
		annual	Variable	booths, exhibits, fairs —attendance and participation	
	Design and implementation of interpretive displays	5 years	River Center	Permanent displays at River Center	
		5 years	River Center, Tribe, WDFW, County parks	Interpretive trail signs	
	Information for recreational boaters	1-5 years	WDFW, County parks, Tribe	Interpretive signs and brochures for boaters at launch sites.	Behavior change or continued stewardship by vessel owners.
	In-class and in-field school programs	ongoing	River Center; Tribe; CCD; County	# of students reached; # of accompanying adults reached	Evaluate student understanding of watershed processes and impacts from actions; participant feedback; teacher feedback

Element h. *A set of criteria that can be used to determine whether the water quality-based goals are being achieved over time and substantial progress is being made*

The criteria can be expressed as indicators and associated interim target values. You can use various indicators to help measure progress. You will want to select indicators that will provide quantitative measurements of progress toward meeting the goals and can be easily communicated to various audiences. It is important to remember that the indicators and associated interim targets will serve as a trigger: If the criteria indicate that you are not making substantial progress, you should consider changing your implementation approach.

The indicators might reflect a water quality condition that can be measured (dissolved oxygen, nitrogen, total suspended solids) or an action-related achievement that can be measured (pounds of trash removed, number of volunteers at the stream cleanup, length of stream corridor revegetated). In other words, the criteria are interim targets in the watershed plan, such as completing certain subtasks that would result in overall pollutant reduction targets. Be sure to distinguish between programmatic indicators that are related to the implementation of your work plan, such as workshops held or brochures mailed, and environmental indicators used to measure progress toward water quality goals, such as phosphorus concentrations or sediment loadings.

Table II-10. Stressors and indicators

Stressor	Measurable indicator
Sediment	Total suspended solids (TSS), turbidity
Eutrophication	Chlorophyll *a*, nitrate/ nitrite/ total phosphorus/ nitrogen, ammonia, dissolved oxygen
Pathogens	Fecal coliforms, *E. coli*
Metals	Copper, lead, zinc
Habitat	Temperature, physical habitat assessed by rapid bioassessment
General water quality	Total dissolved solids (tds), conductivity, pH, oil and grease
Flow	Dry-weather flow, peak flow, flood event frequency
Biology	Diversity and richness indices, biological indices, macroinvertebrates, basic habitat

In the case of scientific indicators, remember that your measurable criteria or indicator links your stressors with your goals. The indicator will tell you whether a BMP is sufficiently addressing the source and achieving your goal. The indicators you selected (e.g., riparian buffers/canopy cover, nutrient concentrations, suspended solids) will serve as the *yardstick* for providing baseline and post-project information. You will also need to select a target value for your project monitoring indicators, so you can measure progress toward your management objectives.

As an example:

Source/Stressor = excessive turbidity of river water.

Indicator (criteria) = nephelometric turbidity units (NTUs), as read on a calibrated digital probe.

Water quality goal = Turbidity must not exceed 5 NTUs over natural background levels when the natural background is 50 NTUs or less, or have more than a 10 percent increase in turbidity when the natural background level is more than 50 NTUs.

The following are some factors you might want to consider when selecting your indicators:

Validity

- *Is the indicator related to your goals and objectives?*
- *Is the indicator appropriate in terms of geographic and temporal scales?*
- *Is the indicator measurable?*

Clarity

- *Is the indicator simple and direct?*
- *Do the stakeholders agree on what will be measured?*
- *Are the methodologies consistent over time?*

Practicality

- *Are adequate data available for immediate use?*
- *Are there any constraints on data collection?*

Clear Direction

- *Does the indicator have clear action implications, depending on whether the change is positive or negative?*

Example: Indicators in the Dungeness WBP (JST 2007)

Criteria for evaluating progress are ... organized by Tribal sub-goal as follows:

Water Quality: Interim criteria have been developed as part of the TMDLs for the Lower Dungeness River/Matriotti Creek and Dungeness Bay. Over the long term, it is the Tribe's intent that bacteria levels are reduced sufficiently that all shellfish beds in the Dungeness area are certifiable. The Detailed Implementation Plan contains tables of the required reductions in fecal coliform concentrations for tributaries to Dungeness Bay, marine sites, Dungeness River, and irrigation ditches to the inner Bay. The interim targets of the Clean Water Workgroup are:

- Achieve target bacteria reductions in the Dungeness River and Matriotti Creek Fecal Coliform Bacteria Total Maximum Daily Load Study (Sergeant 2002)
- For the Dungeness Bay Fecal Coliform Bacteria Daily Load Study (Sergeant 2004), the targets are as follows:
  - Approximately 80% of required reduction by 2010
  - Achievement of standards and restored shellfish harvest by 2012

In addition to the interim targets based on fecal coliform concentrations, it is the goal of the Clean Water Work Group to meet all other water quality standards by 2012 and maintain them thereafter. These include all the parameters required under the Clean Water Act section 106 program.

Element i. A monitoring component to evaluate the effectiveness of the implementation efforts over time

As part of developing your watershed plan, you need to develop a monitoring component to track and evaluate the effectiveness of your implementation efforts using the criteria developed in the previous section.

Monitoring programs can be designed to track progress in meeting tribal water quality goals, which can include load-reduction goals and attaining water quality standards. As in any environmental program, there are significant challenges to overcome. Clear communication between program and monitoring managers is important to specify monitoring objectives that, if achieved, will provide the data necessary to satisfy all relevant management objectives. The selection of monitoring designs, sites, parameters, and sampling frequencies should be driven by the agreed-upon monitoring objectives, although some compromises are usually necessary because of factors such as site accessibility, sample preservation concerns, staffing, logistics, weather and costs. If compromises are made because of constraints, it is important to determine whether the monitoring objectives will still be met with the modified plan. There is always some uncertainty in monitoring efforts, but to knowingly implement a monitoring plan that is fairly certain to fail is a complete waste of time, effort, and resources. Because statistical analysis is usually critical to the interpretation of monitoring results, it is usually wise to consult a statistician during the design of a monitoring program.

Consider a range of objectives like the following when developing your monitoring program:

- ▣ Analyze long-term trends

- ▣ Document changes in management and pollutant source activities in the watershed

- ▣ Measure the performance of specific management practices or implementation sites

Water blessing ceremony on the Winnebago Reservation in Nebraska.

☐ Calibrate or validate models

☐ Fill data gaps in the watershed characterization

☐ Track compliance and enforcement in point sources

☐ Provide data for educating and informing stakeholders

When developing a monitoring design to meet your objectives, it is important to understand how the monitoring data will be used. Ask yourself questions like the following:

☐ What questions are we trying to answer?

☐ What assessment techniques will be used?

☐ What statistical power and precision are needed?

☐ Can we control for the effects of weather and other sources of variation?

☐ Will our monitoring design allow us to attribute changes in water quality to the implementation program?

Remember that you might not see success the very first time that you try new BMPs. That is why the watershed approach includes a feedback loop for continual improvement. Set realistic goals that you can achieve, keep your stakeholders enthusiastic, and make improvements where and when needed!

Example: Monitoring Activities identified in the Dungeness WBP (JST 2007)

In addition to water quality parameters (temperature, turbidity, bacterial, nitrates, metals, etc.), the Tribe and partners have extensive baseline and ongoing monitoring of ecological processes, habitat conditions, and the status of plant and aquatic biological communities. It is the goal of the Tribe and partners to monitor all nine parameters required under the Clean Water Act Section 106 programs by 2012. This will require additional staff training, equipment, laboratory services, data analysis, and preparation of new and updated Quality Assurance Project Plans. Ongoing monitoring and adaptive management is dependent upon adequate funding for these activities.

Regional Review of Watershed-Based Plans

Tribes that would like EPA to review their WBPs should submit their plans to their EPA project officer. The project officer or another EPA staff member will review each WBP to determine whether the plan has the nine minimum elements required in EPA watershed plans. For an example of a WBP checklist for EPA Region 10, see *http://yosemite.epa.gov/R10/ ecocomm.nsf/Watershed+Collaboration/State+Tribal+NPS/$FILE/Tribal-319-Checklist.pdf.* Other EPA regions might use different types of checklists in reviewing WBPs. For more information, contact your Regional Tribal NPS coordinator.

The review is not a regulatory or legal requirement for tribes. The primary purpose of the review is to work with tribes to develop plans that can greatly increase the chances of improving water quality within reservation waters, as well as waters upstream and downstream of reservation boundaries. Tribes may also implement a WBP that was not developed by the tribe. For example, some tribal reservation watersheds might reside in a small portion of the reservation. The remainder of the watershed is on land where a WBP was developed by another organization. In such cases, tribal water staff could submit to EPA the WBP developed by the nontribal group to help determine its impact on reservation waters.

Although EPA strongly recommends developing WBPs, a tribe could still receive funding for projects not part of a WBP, as long as it shows that the project contains many aspects of a watershed planning process and will contribute to improving water quality conditions on a watershed scale. Priorities and requirements for tribal NPS implementation projects in the annual RFPs could change over time, so applicants should read the most recent publication for current information.

First National Tribal NPS Workshop, September 2009, Morongo Reservation, CA.

Clean water is not an expenditure of federal funds; clean water is an investment in the future of our country.

—BUD SHUSTER, U.S. REPRESENTATIVE
QUOTED IN THE WASHINGTON POST, JANUARY 9, 1987

Leveraging Funding Resources

The base and competitive funding provided by EPA through the national NPS pollution control program will not usually be sufficient to address all the polluted runoff and aquatic habitat problems on tribal lands. The tribes having the most success with improving water quality are usually those that are able to leverage other resources, such as assistance from technical service providers, support from other programs interested in water resources, in-kind labor provided by volunteers or students, other departments in the tribe with available funds or means of support such as equipment, or funding from outside sources. This section describes some approaches for acquiring technical, labor, funding, or other resources to support tribal NPS management programs and projects. The topics below start with those resources that are probably closest to the tribe's NPS pollution program and move outward to other related resources.

Tribal CWA 106 Program

Section 106 of the CWA authorizes federal grants to assist state and interstate agencies in administering water pollution control programs. Federally recognized tribes that have applied for and received treatment in a manner similar to a state (TAS) and meet the requirements of CWA section 518(e) can receive CWA section 106 funding. That funding allows tribes to address water quality issues by developing monitoring programs, water quality assessment, standards development, planning, and other activities designed to manage reservation water resources.

Most tribes use staff supported by their CWA section 106 grants to develop their CWA section 319 assessment reports and management program plans. EPA also encourages tribes to use 106 funding to develop WBPs. Section 106 funds can also be used for inventorying nonpoint sources, attending NPS meetings and training, and forming partnerships to address NPS issues. After the CWA section 319 funding is received and the tribal NPS pollution program is established, the CWA section 106 program provides an important source of support for technical and other services, which can be integrated with the nonpoint program into the tribe's overall water resource management agency or program. Section 106 funds focus on

The Upper Sioux Community conducts water resource outreach (EPA Region 5 NPS workshop, 2007).

planning and management activities and may not be used to implement WBPs or for on-the-ground projects. Any pre- and post-project monitoring efforts should be incorporated into 319-funded projects; however, 106 funds can be used for this monitoring if needed. Tribes should contact their EPA section 106 coordinator for further information. Information on the overall section 106 program is at *www.epa.gov/owm/cwfinance/pollutioncontrol.htm*. Information specific to the section 106 tribal guidance is at *www.epa.gov/owm/cwfinance/106tgg07.htm*.

State CWA 319 Programs

Tribes and states are authorized under CWA section 319 to implement approved NPS management programs. Generally, 319 funding supports nonregulatory or regulatory programs for enforcement, technical assistance, education, training, technology transfer, and demonstration projects (per 319(b)(2)(B)). Tribes and states also receive funding under section 319 to support special projects that result in developing or implementing WBPs and other activities that support NPS pollution control, such as stream restoration, public education, septic system repair/replacement, rangeland management, and agricultural practices to control soil erosion.

In many states, tribes may apply for the federal *pass through* CWA section 319 funds disbursed by state water resource agencies. States have been especially interested in funding NPS pollution projects that include multi-stakeholder groups, such as tribes, agricultural agencies, producer groups, and environmental organizations. The match requirements for projects in most states are high—40 percent—but in many cases, more funding is available than that offered by the national tribal NPS program. Because each state develops its own rules regarding CWA section 319 grants, tribes interested in applying for state funding should contact that state's NPS pollution program coordinator. A list of state CWA 319 program administrators and their contact information is posted at *www.epa.gov/nps/state_nps_coord.pdf.*

Meeting the higher state matching support requirements can be challenging, but there are many examples of how to address the need for in-kind and cash support. Below are some approaches to consider:

Barry Tonning

Monument to Waŋbdí Okíčhize (Chief War Eagle), confluence of the Big Sioux and Missouri Rivers.

- **Labor.** Volunteers who work on a project by attending meetings, planning sessions, training events, volunteer monitoring programs, or contributing field work represent value to the project that can be considered in-kind match. Volunteer time can be valued at actual pay rates, including fringe and other costs, or it can be estimated according to similar pay for similar work. General information on the value of volunteers' time is at *www.independentsector.org/programs/research/volunteer_time.html.*

▢ **Equipment Usage.** The value of having a piece of equipment such as a chain saw, tractor, or backhoe can be estimated. Usually, such services are valued on a per-hour basis, at a rate that includes the time of the operator of the equipment. For example, average hourly rates for an operator and a small, medium, or large tractor can be estimated at $30, $40, and $50 or more, respectively. Average rates for a backhoe and operator range from $30 to $60 per hour, depending on the location and equipment. A worker with a chain saw can be valued at $15 to $30 per hour, depending on the worker's skill and efficiency.

▢ **Easements.** Projects that involve temporary or permanent use of a piece of land derive value from the use of that land, and the value can be measured. Valuation of an easement that is being used to address polluted runoff or the degradation of a waterbody can be done by comparing the cost for acquiring a similar easement for commercial or other purposes. For example, the value of an easement along a river or stream that has been granted as part of a project to plant riparian buffer vegetation can be measured in acres and compared to the cost of renting similar acreage on an annual or other basis.

▢ **Funding.** Direct funding or other cash support provided as part of an NPS pollution abatement project is always accounted for in actual terms. For example, if a county, nonprofit organization, or other entity donates a sum of money or directly pays for activities that support a project, that amount can be considered project matching support.

Wetlands Program

What do wetlands have to do with NPS pollution control? Each wetland is a part of a watershed. The actions upstream can affect a wetland through surface or ground water. For example, excessive erosion upstream can fill in a wetland with too much sedimentation. A wetland can also reduce the impact of NPS pollution by taking up nutrients or filtering out sediments, if it is not overwhelmed. Wetland monitoring and assessment might also give you more information on polluted runoff and water quality improvements.

The CWA section 104(b)(3) Wetland Program Development Grants provide eligible applicants an opportunity to improve wetland programs by conducting projects that promote the coordination and acceleration of research, investigations, experiments, training, demonstrations, surveys, and studies relating to the causes, effects, extent, prevention, reduction, and elimination of water pollution. States, tribes, local governments, interstate associations, intertribal consortia, and national nonprofit, nongovernmental organizations are eligible to apply.

The eligible development activities are based in the Four Core Elements of a State/Tribal Wetlands Program found at *www.epa.gov/owow/wetlands/initiative/estp.html*. They include the following:

- Develop and refine a wetland protection program

- Develop a comprehensive monitoring and assessment program

- Improve the effectiveness of compensatory mitigation

- Refine the protection of vulnerable wetlands and aquatic resources

- Complete a wetland restoration demonstration project using a new method, monitor the change in wetland condition and water quality, and incorporate lessons learned into a new wetland program plan

- Use existing wetland data to figure our priorities for wetlands restoration and protection

- Develop an official definition of a *wetland*

- Train staff to monitor wetland condition and water quality associated with wetlands

For more information on the Wetland Program Development Grants, EPA Regional priorities and contacts, and tools for monitoring wetlands, go to the EPA Web site at *www.epa.gov/wetlands.*

Solid Waste

Solid waste is detrimental to reservation water resources and community health. Runoff from illegal dump sites can contain hazardous chemicals that can pollute ground water and drinking water. This is especially detrimental in tribal communities that depend on wells as a main source of freshwater. Illegal dumping on the sides of streams can destroy fish and wildlife habitat. Many tribes depend on these habitats to foster wildlife that play important environmental, spiritual, and economic roles in community life. Even backyard waste burning, which seems as if it would not directly affect water quality, can be damaging to tribal water sources. Runoff through burn sites can pick up chemicals and carry them to water sources; in addition, outfall from the burning process is released into the air and can be a factor in the contamination process.

Solid waste management program activities that contribute to water quality restoration can be funded by CWA section 319 base and competitive grants. Funds may cover a range of activities, including dump site cleanup, collection platform construction, solid waste management training, post-project water quality monitoring, and outreach activities within the local communities. Supplies and vehicle procurement can be funded under the section 319 program, but federal grant programs generally do not allow funds to be used for program or facility operation and maintenance. For more information, check with your EPA Regional representative on allowable activities. For more general information on tribal solid waste management programs, refer to EPA's Tribal Solid Waste Management page at *www.epa.gov/epawaste/wycd/tribal*. Another resource is the *Tribal Decision Maker's Guide* at *www.epa.gov/osw/wycd/tribal/tribalguide.htm*. The guide provides an overview of solid waste management program development and includes information on solid waste planning, regulations, collection, disposal, recycling, and education, along with tribal case studies.

Drinking Water and Clean Water State Revolving Fund Programs

Clean Water State Revolving Funds

The Clean Water State Revolving Funds (CWSRF) program provides low-interest loans that can spread project costs over a long term repayment period. Repayments are then cycled back into the fund and used to pay for additional clean water projects. Although the majority of the fund has traditionally been used to finance CWA section 212 projects, such as wastewater treatment and collection facilities, financing is also available for NPS projects under the authorities of CWA sections 319 and 320. Eligible NPS problems to address include the following:

- Agriculture runoff
- Leaking on-site septic systems and underground storage tanks
- Urban NPS pollution, including stormwater runoff and hazardous waste contamination (for more information see *www.epa.gov/brownfields*)
- Forestry issues
- Hydromodification
- Estuary protection
- Atmospheric deposition
- Runoff from closed landfills and abandoned mines
- Source water protection

Public or private entities, including tribes, local governments, watershed groups, agricultural organizations, farmers, and other eligible borrowers may apply for CWSRF loan funding. Each state controls its own CWSRF program and determines project eligibility requirements for the loans and sets interest rates. To be eligible under the broad authority of CWA sections 319 and 320, a project must help implement the state's NPS management plan (CWA section 319) or be consistent with actions and priorities contained in a National Estuary Program Comprehensive Conservation Management Plan (CWA section 320). In some instances, Congressional appropriations allow some portion of these funds to be awarded as a direct grant. For more information, contact CWSRF staff. For a list of CWSRF contacts, go to *www.epa.gov/owm/cwfinance/cwsrf/cwnims*.

Drinking Water State Revolving Fund: Source Water Protection Funding

Like the CWSRF, the Drinking Water State Revolving Fund (DWSRF) Set-Aside program also provides low-interest loans to communities for addressing threats to drinking water sources.

Tribes in watersheds that have drinking water treatment plants might want to consider the Source Water Protection Program as a possible source of support for projects that protect the quality of surface or ground water. Under the federal Safe Drinking Water Act, water utilities with at least 15 service connections or that regularly serve at least 25 or more people per day must have a source water assessment, which includes delineation of the source water protection area (the portion of a watershed or ground water recharge area that might contribute water and possibly pollutants to the water supply), identification of all significant potential sources of drinking water contamination within the protection area, a determination of the water supply's susceptibility to contamination from those sources, and making the source water assessment results available to the public. Tribes are not required to develop source water assessments, but EPA Regions with direct implementation responsibility for tribal public water systems are completing such assessments for tribes in their Regions. Tribal publicly and privately owned and nonprofit, non-community water systems are eligible to receive DWSRF funding from the state to address source water protection efforts. Tribes can contact their DWSRF loan fund managers to check on eligibility requirements and other details. For more information, see *www.epa.gov/safewater/dwsrf*.

Source water protection measures that address the potential contaminants identified by the assessment are largely the responsibility of local drinking water utilities. The information in the assessment often represents a significant source of data regarding possible nonpoint sources of pollution in the area, and public interest in protecting the drinking water supply can provide a very powerful incentive for addressing those sources. Drinking water utilities often support a variety of source water protection structural

Restoration project field trip, Region 6 Tribal NPS Workshop.

and nonstructural practices, which can include regulatory and nonregulatory land use controls, waste site cleanups, acquiring or planting buffer areas, purchasing development rights, developing design standards, good housekeeping practices, public education, septic system improvement projects, and measures targeted at animal waste, pesticide application, and fertilizer use. Nearly all these have been identified as practices that control nonpoint sources of water pollution.

General Assistance Program (GAP) Funding

The primary purpose of the GAP is to support the development of a core tribal environmental protection program. To achieve this goal, the tribes, with EPA's assistance, use GAP to do the following:

- Identify baseline environmental needs to build capacity to administer an environmental program or develop a tribal environmental program that is tailored to individual tribal needs

- Establish the administrative, legal, technical, and enforcement capability of tribes to develop and implement a tribal environmental program, including the capacity to manage EPA-delegated programs

- Foster compliance with federal environmental statutes by developing appropriate tribal environmental programs, ordinances, and public education and outreach programs

- Establish a tribal communications capability to work with federal, state, local, tribal and other environmental officials

- Establish the tribal capacity to develop and implement management programs through program-specific assistance

Although GAP funds may be used to develop program capacity, they may not be used for program implementation. GAP guidance documents are at *www.epa.gov/indian/gap.htm.*

Humankind has not woven the web of life.
We are but one thread within it.
Whatever we do to the web, we do to ourselves.
All things are bound together. All things connect.

—CHIEF SEATTLE, 1854

Environmental Quality Incentive Program

The Environmental Quality Incentives Program (EQIP) administered by the U.S. Department of Agriculture is a voluntary program that provides financial and technical assistance to farmers and ranchers who face threats to soil, water, air, and related natural resources on their land. Through EQIP, USDA's Natural Resource Conservation Service provides financial incentives to producers to promote agricultural production and environmental quality as compatible goals, optimize environmental benefits, and help farmers and ranchers meet federal, state, tribal, and local environmental regulations. Tribal owners of land in agricultural production or members who are engaged in livestock or agricultural production on eligible land may participate in EQIP. National EQIP priorities include the following:

1. Reductions of NPS pollution, such as nutrients, sediment, pesticides, or excess salinity, in impaired watersheds as required by TMDLs, where applicable, as well as the reduction of ground water contamination and reduction of point sources such as contamination from confined animal feeding operations

2. Conservation of ground water and surface water resources

3. Reduction of emissions, such as particulate matter, nitrogen oxides, volatile organic compounds, and ozone precursors and depleters, that contribute to air quality impairment violations of the National Ambient Air Quality Standards

4. Reduction in soil erosion and sedimentation from agricultural land

5. Promotion of at-risk species habitat conservation

For more information on EQIP and applying for funds, see *www.nrcs.usda.gov/PROGRAMS/EQIP.*

Other Programs

In addition to EQIP funding, many tribes partner with the U.S. Fish and Wildlife Service, Bureau of Land Management, U.S. Forest Service and other federal and state agencies in developing watershed protection programs, technical assistance and education projects, cost-share program priorities, and other activities. Those partnerships are extremely helpful to tribes, which can benefit directly from technical and financial support, and to the partner groups, which often seek tribal input in addressing polluted runoff issues in watersheds where tribal lands might be. Pooling resources usually involves coordination of staff activities under a formal or, in most cases, an informal plan. That can include efforts to conduct water quality monitoring activities, sharing data to produce a watershed assessment, developing technical assistance programs for farmers or ranchers, promoting cost-share signups for livestock-exclusion stream fencing, septic system inspection/repair projects, stabilizing unpaved roads, restoring stream corridors, removing invasive species, and other efforts. In general, such funding sources seek to support projects that are clearly explained, have a measurable water quality or public health benefit, draw support from several programs or partners, and are sponsored by organizations with a successful track record.

PART III

Additional Resources for Tribes

List of Contacts

For current contacts for EPA tribal NPS coordinators and state NPS coordinators, as well as NPS coordinators from each state and territory, see *www.epa.gov/nps/contacts.html*. That site is updated often and provides links to state NPS programs. For general EPA tribal contacts and Regional tribal program information, see *www.epa.gov/tribal/contactinfo*.

General EPA tribal nonpoint source contact information

Headquarters	
Tribal Coordinator, USEPA	
1200 Pennsylvania Avenue, NW (MC 4503T)	
Washington, DC 20460	
202-566-1155	
www.epa.gov/owow/nps/tribal/index.html	

Region 1 (CT, ME, MA, NH, RI, VT)	**Region 6 (AR, LA, NM, OK, TX)**
Tribal Coordinator	Tribal Coordinator
USEPA - Region 1	USEPA - Region 6
5 Post Office Square, Suite 100 (MC: OEP06-1)	1445 Ross Avenue, Suite 1200 (MC: 6WQ-AT)
Boston, MA 02109-3912	Dallas, TX 75202
617-918-1840	214-665-6684
www.epa.gov/region1/govt/tribes/index.html	*www.epa.gov/region6/water/at/tribal/index.htm*
Region 2 (NJ, NY, PR, VI)	**Region 7 (IA, KS, MO, NE)**
Tribal Coordinator	Tribal Coordinator
USEPA - Region 2	USEPA - Region 7
290 Broadway Avenue, 24th floor (MC: DEPP: WPB)	901 North 5th Street
New York, NY 10007	Kansas City, KS 66101
212-637-3788	913-551-7003
www.epa.gov/region02/nations/intro.htm	Toll-free: 1-800-223-0425
	www.epa.gov/region07/tribal/index.htm
Region 3 (DE, DC, MD, PA, VA, WV)	**Region 8 (CO, MT, ND, SD, UT, WY)**
Nonpoint Source Coordinator	Tribal Coordinator
(No federally recognized tribes have been registered in Region 3.)	USEPA - Region 8
USEPA - Region 3	1595 Wynkoop Street (MC EPR-EP)
1650 Arch Street	Denver, CO 80202
Philadelphia, PA 19103	303-312-6895
215-814-5753	*www.epa.gov/region8/tribes/contacts.html*
www.epa.gov/reg3wapd/nps/index.htm	

General EPA tribal nonpoint source contact information *(continued)*

Region 4 (AL, FL, GA, KY, MS, NC, SC, TN)	Region 9 (AZ, CA, HI, NV, AS, GU)
Tribal Coordinator USEPA - Region 4 Atlanta Federal Center 61 Forsyth Street, SW Atlanta, GA 30303 404-562-9451 *www.epa.gov/region4/indian/contacts.htm*	Tribal Coordinator USEPA - Region 9 75 Hawthorne Street (MC WTR-10) San Francisco, CA 94105 415-972-3402 *www.epa.gov/region09/water/tribal/* *tribal-cwa.html#nps*
Region 5 (IL, IN, MI, MN, OH, WI)	Region 10 (AK, ID, OR, WA)
Tribal Coordinator USEPA - Region 5 77 West Jackson Boulevard (MC WS-15J) Chicago, IL 60604 312-353-2000 *www.epa.gov/region5/water/wshednps/topic_nps.htm*	Tribal Coordinator USEPA - Region 10 1200 Sixth Avenue, Suite 900 (MC OWW-137) Seattle, WA 98101 206-553-1050 *yosemite.epa.gov/R10/ecocomm.nsf/* *Watershed+Collaboration/* *State+Tribal+NPS#Tribes%20Section*

EPA Regional Offices

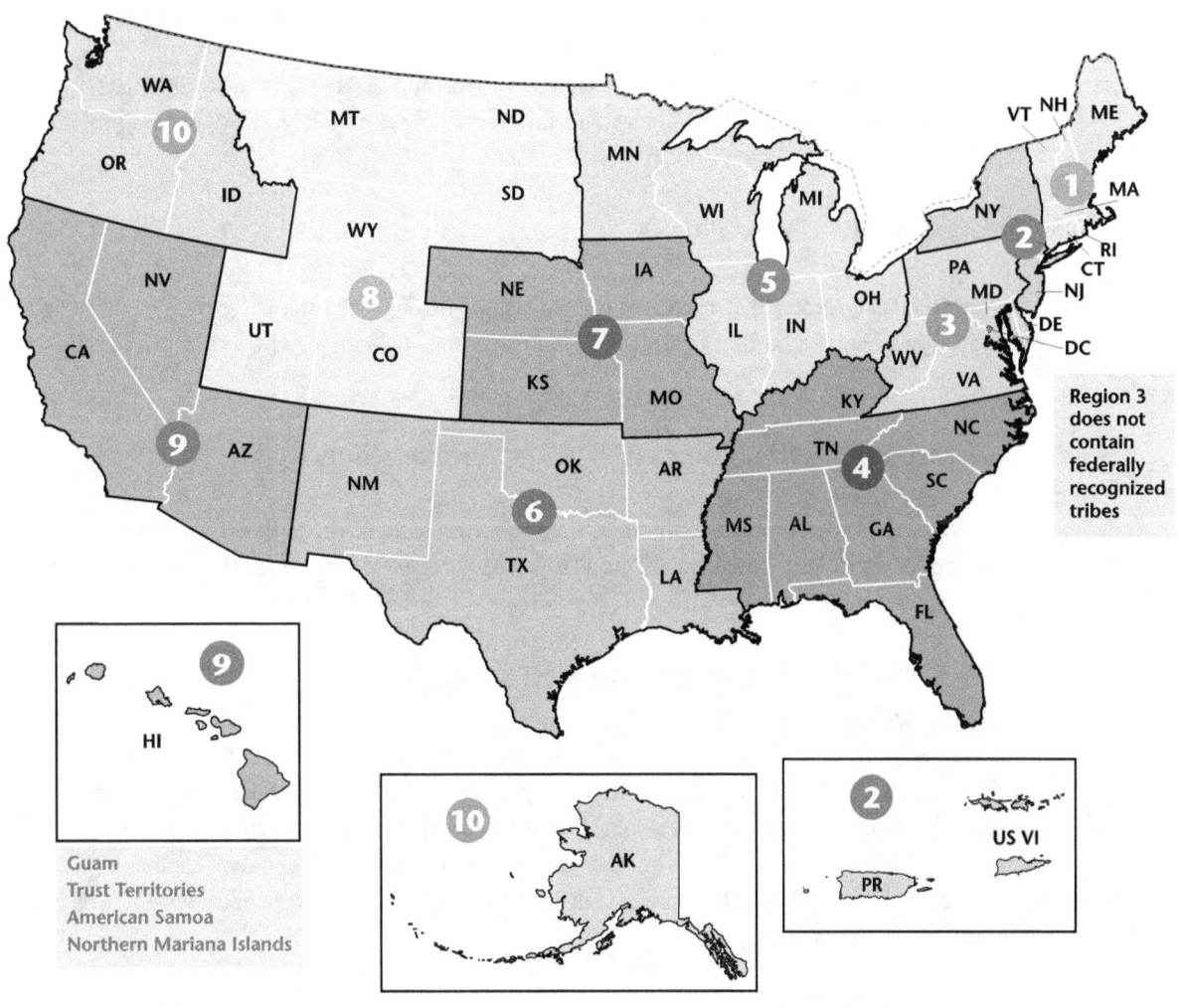

Online Tools

EPA offers a host of online tools to help with the process of conducting a watershed assessment. To develop the scope of a watershed planning effort or waterbody restoration project, the watershed must be characterized. The characterization process allows for (1) defining of issues of concern, the key pollutants, and the sources of pollution; (2) assessing watershed geographic conditions, economic activities, and discharges that could affect water pollution; and (3) selecting target restoration efforts. The following Web sites provide watershed characterization information.

Surf Your Watershed
www.epa.gov/surf
Watersheds and watershed resources can be located through an interactive map or through drop-down lists. Eight-digit Hydrologic Unit Codes (HUCs) can also be found through this resource.

Assessment TMDL Tracking and Implementation System (ATTAINS)
www.epa.gov/waters/ir
ATTAIN's tables and charts summarize state-reported data for the nation as a whole, individual states, individual waters, and the 10 EPA Regions. It provides the *full story* of assessed waters that are impaired, are being restored, or have been restored.

EPA's Watershed Assessment and Tracking Environmental Results (WATERS)
www.epa.gov/water
The EPA Office of Water manages numerous programs that collect and store water quality-related data in separate databases. WATERS is an integrated information system that generates reports on the nation's surface waters from multiple, independent databases (e.g., Water Quality Standards Database, STORET, ATTAINS, Permit Compliance System, Clean Watersheds Needs Survey).

Storage and Retrieval (STORET)
www.epa.gov/storet
STORET is EPA's main repository of water quality monitoring data, including chemical, biological, and physical data. It contains water quality information from a variety of organizations across the country, from small volunteer watershed groups to state and federal environmental agencies. Data stored in STORET are accessible to anyone with a Web browser and Internet access, and query results can be e-mailed to the user in an Excel format.

Enforcement and Compliance History Online (ECHO)

www.epa-echo.gov/echo

ECHO focuses on facility compliance and EPA/state enforcement of environmental regulations. Roughly 800,000 regulated facilities under the following environmental statutes and regulations are included in ECHO: Clean Air Act Stationary Source Program, Clean Water Act, National Pollutant Discharge Elimination System, and Resource Conservation and Recovery Act. Through ECHO, site visitors can find inspection, violation, enforcement action, informal enforcement action, and penalty information about facilities for the past three years.

For introductory instructional materials on how to use the above EPA Internet tools, visit *www.epa.gov/watershed/wacademy/epatools* and download the course materials for Key EPA Internet Tools for Watershed Management.

GIS Tools

MyEnvironment (Formerly, Window to My Environment)

www.epa.gov/myenvironment

MyEnvironment is designed to improve access to useful community-based environmental information. MyEnvironment represents a concerted effort to develop a *geographic portal* for integrating that environmental information by local geography to help answer common questions, examine critical problems, and discover potential solutions for environmental protection and human health issues.

MyEnvironment's features include

- ☐ Interactive Map: Shows the location of regulated facilities, monitoring sites, waterbodies, population density, perspective topographic views and more with hotlinks to state/federal information about these items of interest.

- ☐ Your Window: Provides selected geographic statistics about the area of interest, including estimated population, county/urban area designations, local watersheds/waterbodies, and so forth.

- ☐ Your Environment: Links to information from federal, state, and local partners on environmental issues like air and water quality, watershed health, Superfund sites, fish advisories, impaired waters, as well as local services working to protect the environment in your area.

Tribal Windows to the Environment

http://oaspub.epa.gov/tims/twe.html

The American Indian Environmental Office (AIEO) coordinates and integrates all the tribal programs at EPA. To track the progress that these programs are making toward protecting the environment and public health in Indian Country, AIEO has linked the various regulatory and environmental monitoring databases of EPA into a single *window to the environment* for tribes. The AIEO windows are essentially the same as the popular "MyEnvironment."

EnviroMapper for Water

www.epa.gov/waters/enviromapper

EnviroMapper for Water is a Web-based geographic information system (GIS) application that dynamically displays water quality and other environmental information about bodies of water in the United States. This interactive tool allows you to create customized maps that portray surface waters along with a collection of water quality-related data from the national level down to community level. Enviromapper for Water provides the ability to:

- Geographically display a variety of EPA water program data (e.g., water quality standards, ATTAINS, and STORET)

- Pan, zoom, label, and print maps

- Link to water program Web reports after identifying specific features of interest

- Generate specific water quality-related reports based on an area of interest

Watershed Plan Builder

www.epa.gov/owow/watershedplanning

The Watershed Plan Builder is designed for users who are just beginning to develop a watershed plan, are in the process of developing a watershed plan, or are updating an existing plan. The Watershed Plan Builder provides a step-by-step process to begin developing a watershed plan. Through a series of questions related to eight categories such as location, pollutants, and stakeholders, the Plan Builder collects information about your specific watershed area and produces a customized watershed outline. This outline can be used as a roadmap for the watershed planning process to create a comprehensive watershed plan. Specific instructions on how to develop a watershed plan can be found in the section devoted to the watershed planning process.

Watershed Central and the Watershed Wiki

www.epa.gov/watershedcentral

EPA has a new Web site called Watershed Central to help watershed organizations and others find key information they need to implement watershed management projects. The primary purpose of the Watershed Central Web site is to make it easy for organizations to find the information that they need to help protect and restore their water resources.

Watershed Central helps users find environmental data, watershed models, nearby local organizations, guidance documents, and other information depending on the task at hand. It also contains links to watershed technical resources, funding sources, mapping applications, and information specific to named watersheds. The site includes a Watershed Central Wiki for collaboration and information sharing. EPA encourages watershed practitioners to use the new Watershed Central Wiki to share tools, scientific findings, expertise, and local approaches to watershed management. Watershed Central not only links to EPA Web resources, but also links to other valuable funding, guidance, and tools on the Web sites of state, tribal, and federal partners, universities, and nonprofit organizations.

Tools of the Trade

www.epa.gov/watershed/wacademy/epatools
Tools of the Trade are publicly available, GIS-based tools to help with effective watershed management.

Additional Internet Resources

EPA

Information

Tribal Nonpoint Source Information
www.epa.gov/nps/tribal

Section 319 Information
www.epa.gov/nps/cwact.html

Tribal Portal Federal Funding Eligibility
www.epa.gov/tribalportal/laws/tas.htm

Treatment in the Same Manner as a State Information
www.epa.gov/tribalportal/laws/tas.htm

Tribal Portal
www.epa.gov/tribalportal

American Indian Tribal Portal: Where You Live
www.epa.gov/tribalportal/whereyoulive

Clean Water Act Tribal Training

www.epa.gov/water/tribaltraining

Nonpoint Source Success Story Web Site

www.epa.gov/nps/Success319

This Web Site features stories about primarily nonpoint source-impaired waterbodies where restoration efforts have led to documented water quality improvements.

Healthy Watersheds Initiative

www.epa.gov/healthywatersheds

The objective of the federal CWA is to "restore and maintain the chemical, physical, and biological integrity of the nation's waters." While other EPA programs focus on restoring impaired waters, the Healthy Watersheds Initiative augments the watershed approach with proactive, holistic aquatic ecosystem conservation and protection. The Healthy Watersheds Initiative includes both assessment and management approaches that encourage tribes, states, local governments, watershed organizations, and others to take a strategic, systems approach to conserve healthy components of watersheds and, therefore, avoid additional water quality impairments in the future.

Green Infrastructure

www.epa.gov/greeninfrastructure

Green infrastructure is an approach to wet weather management that is cost-effective, sustainable, and environmentally friendly. Green Infrastructure management approaches and technologies infiltrate, evapotranspire, capture, and reuse stormwater to maintain or restore natural hydrologies.

EPA Grant Finding Resources

Clean Water Indian Set-Aside Program

www.epa.gov/owm/mab/indian/cwisa.htm

Coastal Program

www.epa.gov/owow/estuaries

Drinking Water State Revolving Fund Indian Set-Aside Program

www.epa.gov/safewater/dwsrf/allotments/tribes/index.html

Federal Funding Database

www.epa.gov/epawaste/wycd/tribal/finance.htm

Grant Writing Resources

www.epa.gov/epawaste/wycd/tribal/pdftxt/grant.pdf

Guide to Federal Grant Resources for Community Organizations, Tribal Organizations, and Tribal Governments

www.epa.gov/epahome/grants.htm

Tribal Portal Grants and Funding

www.epa.gov/tribal/grantsandfunding

Tribal Resource Directory for Drinking Water and Wastewater Treatment

www.epa.gov/owm/mab/indian/tribal-resource-directory.htm
PDF version: www.epa.gov/owm/mab/indian/pdfs/tribal-complete.pdf

Watershed Funding

www.epa.gov/owow/funding.html

Wetlands Protection Programs

www.epa.gov/owow/wetlands

Federal Grant Finding Programs

Federal Grants Database

www.grants.gov

U.S. Bureau of Reclamation

www.usbr.gov

U.S. Department of Agriculture Farm Service Agency Programs

www.fsa.usda.gov/pas/default.asp

U.S. Department of Agriculture Natural Resource Conservation Service Programs

www.nrcs.usda.gov/programs

U.S. Department of Agriculture Watershed Program

www.nrcs.usda.gov/programs/watershed

U.S. Department of Health and Human Services Indian Health Services Wastewater Program

www.ihs.gov/nonmedicalprograms/dsfc

U.S. Department of the Interior Abandoned Mines Program

www.osmre.gov/aml/aml.shtm

U.S. Fish and Wildlife Partners for Fish and Wildlife Program

www.fws.gov/partners

Nonfederal Grant Finding Programs

Rural Community Assistance Partners

www.rcap.org

The Environmental Finance Center

www.efc.umd.edu/host.html

Other

Informational

Catalogue of Federal Domestic Assistance

www.cfda.gov

Conservation Technology Information Center

www.conservationinformation.org

Federal Register

www.gpoaccess.gov/fr

Indian Health Service

www.dsfc.ihs.gov

Climate Change Information

EPA's Climate Change Web site

www.epa.gov/climatechange

EPA's Office of Water Climate Change Web site

www.epa.gov/water/climatechange

EPA Watershed Academy — Climate Change and Water Training Module

This module was designed to educate water program managers, as well as the general public, on the expected effects of climate change on water resources and water programs
www.epa.gov/watertrain/climate_water/

ENERGY STAR

www.energystar.gov

WaterSense

www.epa.gov/watersense

Low Impact Development

www.epa.gov/nps/lid

Intergovernmental Panel on Climate Change (IPCC)

www.ipcc.ch

U.S. Climate Change Science Program

www.globalchange.gov

National Oceanic and Atmospheric Administration's Climate Program Office

The Regional Integrated Sciences and Assessments provides links to university networks developing information at the regional level.
www.climate.noaa.gov/index.jsp?pg=./cpo_pa/cpo_pa_index.jsp&pa=risa

A Few Web Links to Notable Documents on Climate Change and Ecosystems

Preliminary Review of Adaptation Options for Climate-Sensitive Ecosystems and Resources, June 2008.
www.globalchange.gov/publications/reports/scientific-assessments/saps/sap4-4
This report from the U.S. Climate Change Science Program focuses on adaptation options for climate-sensitive ecosystems and resources on federally owned and managed lands.

Climate Ready Estuaries: Synthesis of Adaptation Options for Coastal Areas, 2009.
www.epa.gov/cre/downloads/CRE_Synthesis_1.09.pdf
This guide provides a brief introduction to key physical effects of climate change on estuaries and a review of on-the-ground adaptation options available to coastal managers to reduce their systems' vulnerability to climate change impacts.

Synthesis Report from the International Scientific Congress, Climate Change: Global Risks, Challenges & Decisions, March 2009.
http://climatecongress.ku.dk
The Climate Congress was organized to update the state of scientific knowledge regarding climate change since the 2007 IPCC Report, in preparation for the United Nations' Conference on Climate Change to be held in Denmark in December 2009.

Intergovernmental Panel on Climate Change Fourth Assessment Report: Climate Change 2007.
www.ipcc.ch/publications_and_data/publications_and_data_reports.htm
Of particular note for ecosystems, see IPCC Working Group II, Climate Change Impacts, Adaptation, and Vulnerability.

Global Climate Change Impacts in the United States, Thomas R. Karl, Jerry M. Melillo, and Thomas C. Peterson (eds.). Cambridge University Press, 2009.
www.globalchange.gov/usimpacts
This report summarizes the science of climate change and the impacts of climate change on the United States, now and in the future. It is largely based on results of the U.S. Global Change Research Program, and integrates those results with related research from around the world. This report discusses climate-related impacts for various societal and environmental sectors and regions across the nation. It is an authoritative scientific report written in plain language, with the goal of better informing public and private decision making at all levels.

Best Management Practices Implementation Appendix
www.waterquality.utah.gov/TMDL/Virgin_River_Watershed_Implementation_Appendix.pdf
This appendix is a manual designed to assist landowners, managers, and technicians in adopting effective and appropriate practices to reduce NPS pollutants entering streams and watercourses. Practices are defined as actions taken by a landowner or manager to reduce pollutant loads from nonpoint sources. In general, practices described in this manual are meant to be implemented in areas immediately adjacent to the stream channel or waterbody. However, many of the treatments can be used effectively in uplands and other areas. This document is a great resource to view photos of various BMPs, and it includes additional information on BMP purpose, pollutants addressed, load reduction potential, and expected maintenance.

Another widely used source that is applicable to agricultural areas is the USDA-NRCS *Field Office Technical Guide: www.nrcs.usda.gov/technical/efotg.*

Acronyms and Abbreviations

°C	degrees Celsius
°F	degrees Fahrenheit
µg/L	micrograms per liter
ANSI/ASQC	American National Standards Institute
ASQC	American Society for Quality Control
ATTAINS	Assessment TMDL Tracking and Implementation System
BIA	Bureau of Indian Affairs
BMP	best management practice
CFR	*Code of Federal Regulations*
CFS	cubic feet per second
CRP	Conservation Reserve Program
CWA	Clean Water Act
CWSRF	Clean Water State Revolving Fund
DNR	Department of Natural Resources
DO	dissolved oxygen
DWSRF	Drinking Water State Revolving Fund
E. coli	*Escherichia coli*
e.g.	for example
ECBI	Eastern Cherokee Band of Indians
ECHO	Enforcement and Compliance History Online
EPA	U.S. Environmental Protection Agency
EQIP	Environmental Quality Incentives Program
FCB	fecal coliform bacteria
FTE	full-time equivalent (staff)
GAP	General Assistance Program
GIS	geographic information system
HUC	Hydrologic Unit Code
i.e.	in other words; that is
MDNR	Minnesota Department of Natural Resources
mg/L	milligrams per liter

MNDOT	Minnesota Department of Transportation
MPCA	Minnesota Pollution Control Agency
N	nitrogen
NA	not applicable
NPDES	National Pollutant Discharge Elimination System
NPS	nonpoint source (pollution)
NRCS	Natural Resources Conservation Service
NTU	nephelometeric turbidity unit
O&M	operation and maintenance
P	phosphorus
pH	co-logarithm of the activity of dissolved hydrogen ions
PPG	Performance Partnership Grant
QA	quality assurance
QA/QC	quality assurance/quality control
QAPP	quality assurance project plan
RCBLSC	Red Cliff Band of Lake Superior Chippewa
RFP	request for proposal
RLBCI	Red Lake Band of Chippewa Indians
RWQCB	Regional Water Quality Control Board
SI	Stressor Identification
STEPL	Spreadsheet Tool for Estimating Pollutant Load
STORET	STOrage and RETrieval System
su	standard units
SWAP	Source Water Assessment and Protection
SWCD	Soil and Water Conservation District
SYBCI	Santa Ynez Band of Chumash Indians
TAS	Treatment in the same manner as a state
TMDL	total maximum daily load
TSI	Trophic Status Index
TSS	total suspended solids
USACE	U.S. Army Corps of Engineers

USDA	U.S. Department of Agriculture
USFW	U.S. Fish and Wildlife (Service)
USGS	U.S. Geological Survey
VSAP	Visual Stream Assessment Protocol
WATERS	Watershed Assessment and Tracking Environmental Results
WBP	watershed-based plan
WisCAP	Wisconsin Community Action Program
WPCP	Water Pollution Control Program
WQS	water quality standards/water quality specialist

Glossary

Antidegradation – A federal water quality requirement prohibiting deterioration where pollution levels are above the legal limit.

Base funding – $30,000 or $50,000 depending on size of reservation.

Beneficial uses – Designations made by states or tribes regarding how a particular waterbody is expected to be used and for what it is to be managed. Examples: cold water fishery, drinking, swimming.

Best management practices (BMPs) – Practices, measures, or actions that are commonly recommended to prevent, reduce, or mitigate pollution from nonpoint sources.

Competitive funding – Funding allocated to projects as established through a national request for proposals.

Cultural issues – Knowledge, belief, behavior, or set of shared attitudes, values, goals, and practices of a specific group. For Native American cultures, some attributes to consider: respect for the natural world, spirituality, elders and children, clans and kinship, leadership and decision-making, history, governance structures, protocols, and laws.

CWA section 104(b)(3) – A granting authority under which awards are made to state water pollution control agencies, interstate agencies, other public or nonprofit private agencies, institutions, organizations, and individuals, for the purpose of research, investigations, experiments, training, demonstrations, surveys, and studies relating to the causes, effects, extent, prevention, reduction, and elimination of pollution.

CWA section 106 – A granting authority under which awards are made to states, tribes, and interstate agencies to assist them in administering programs for the prevention, reduction, and elimination of pollution, including enforcement. May fund a wide range of water quality activities, including water quality planning and assessments, development of water quality standards, ambient monitoring, development of total maximum daily loads, issuing permits, ground water and wetland protection, and NPS control activities (including nonpoint source assessment and management plans).

CWA section 303(d) – Section under which states, territories, and authorized tribes are required to develop lists of impaired waters that do not meet water quality standards or use designations that have been set for them. The section requires establishing priority rankings for waters on the lists.

CWA section 305(b) – Requires states and territories to report every two years on the water quality and use designations of all navigable waters, surface waters, and ground water and impacts from both point and nonpoint sources of pollution. (Tribes are not required to submit 305(b) reports.)

CWA section 401 – Water quality certification needed to show that an applicant for a federal license or permit to conduct activities that could result in any discharge into the navigable waters will provide to the licensing or permitting agency a certification from the state/tribe or interstate agency having jurisdiction over those waters that any such discharge will comply with the applicable water quality regulations and effluent limits.

CWA section 404 – Establishes permits for disposal of dredged or fill material at specified disposal sites into navigable waters, including notice and opportunity for public hearings.

CWA section 518 – Establishes that Indian tribes will be treated as states for the purposes of title II (grants for treatment works) and sections 104, 106, 303, 305, 308, 309, 314, 319, 401, 402, and 404.

E. coli (Escherichia coli) – A gram negative bacterium that is commonly found in the lower intestine of warm-blooded animals.

Fee lands – Land parcels that are owned by nontribal individuals or entities and are within the reservation boundaries.

Green space – Open spaces that serve as natural assets for the community, such as parkland or naturalized areas necessary for the protection of the waterbody.

Hydrologic Unit Code (HUC) – A 2- to 12- digit number assigned by the U.S. Geological Survey as part of its surface waterbody classification system.

Impaired waters – Those waters that do not meet water quality standards for one or more pollutants and for which the use designation therefore cannot be fulfilled.

Impairments – The kinds of pollutants that creates a condition, by means of amount or type, where water quality standards are exceeded.

Indicator – Entity, process, or community whose characteristics show the presence of specific environmental conditions.

Inputs – When referring to pollution sources, identifying where they come from.

Intertribal consortium – Consortium to promote cooperative work among tribes; partnership between two or more tribes that is authorized by the governing bodies of those tribes.

Karst – A geologic formation of irregular limestone deposits with sinks, underground streams, and caverns.

Landscape scale – Traits, patterns, and structure of a specific geographic area, including its biological composition, its physical environment, and its anthropogenic or social patterns. A geophysical space where interacting ecosystems are grouped and have similar attributes.

Legislative conditions – The requirements as found in the authorizing legislation for a program.

Milestones – Key dates when certain measurable outcomes are expected.

Mitigation – Measures that are taken to reduce adverse effects on the environment and can provide a method of compensation for unavoidable impacts.

Narrative criteria – Statements that describe the desired water quality goal, such as waters being *free from* pollutants or substances that can harm people and fish; an approach used for pollutants for which numeric criteria are difficult to establish because of inherent subjectivity.

Nonpoint source pollution – Pollution not discharged from a point source. This generally consists of pollution from diffuse sources (i.e., without a single point of origin or not introduced into a receiving stream from a specific outlet). The pollutants are generally carried off the land as a result of precipitation events (rainfall, snowmelt).

Nonprofit/nongovernmental organizations – Sometimes seen as NPO or NGO. A group organized for purposes other than generating profit and in which no part of the organization's income is distributed to its members, directors, or officers. This is established at the time of formation, and only approved activities under this designation are allowed; no official governmental representatives are governing members.

Numeric criteria – A number standard for limiting a particular pollutant that protects a specific use designation; can be load- or concentration-based.

Outcomes – The conditions that result from an action.

Outputs – The results of activities undertaken to achieve the outcomes.

Partnership – A cooperative relationship between people or groups that agree to share responsibility for achieving some specific goal.

Performance goals – Numerical or statistically measured achievements against a target.

Permitting/Enforcement authority – Any entity at the local, state, or national level that has the capability to execute a permit or take an enforcement action.

Point source – A stationary location or fixed facility from which pollutants are discharged through a conveyance system; any single identifiable source of pollution, such as a pipe, ditch, ship, ore pit, or factory smokestack.

Public participation – A principle or practice that seeks out and facilitates the involvement of those potentially affected by or interested in a decision. The full range of actions employed to engage people in current or proposed activities. Implies that the public's contribution will influence the decision-making process.

Remediation – Cleanup, restoration, or other methods used to ensure that the location will be able to fully function from an ecological and human health perspective.

Restoration/rehabilitation – Measures taken to return a site to a previous condition.

Riparian areas – Areas adjacent to rivers and streams with a differing density, diversity, and productivity of plant and animal species relative to nearby uplands.

Sampling design – The method employed to collect adequate and appropriate data to support accurate analysis. See *www.epa.gov/quality/qksampl.html*.

Schedule – Weekly, monthly, seasonal, or quarterly array of activities or actions needed to carry out a work plan.

Section 319 base funding – The funding that is provided by CWA section 319 as the core funding to support the 319 program with the broadest ability to accomplish program goals and activities.

Spatial context (units) – The areal extent or scale at which analysis is given or information collected. A description of the nature of the physical setting in which activities occur.

Stakeholder – Any organization, governmental entity, or individual that has a stake in or could be affected by a given approach to environmental regulation, pollution prevention, energy conservation, and the like.

Stressor – Physical, chemical, or biological entity that can induce adverse effects on ecosystems or human health.

Subbasins – Usually, catchments that are smaller than a 12-digit Hydrologic Unit Code and are determined to be a more useful planning and implementation unit for water resources protection or can be contained within the jurisdiction of the tribe.

References

ANSI/ASQ (American National Standards Institute/American Society for Quality). 1994. *Specifications and Guidelines for Quality Systems for Environmental Data Collection and Environmental Technology Programs.* ANSI/ASQ E4-1994. American Society for Quality, Milwaukee, WI.

Burkett, V., J.O. Codignotto, D.L. Forbes, N. Mimura, R.J. Beamish, and V. Ittekkot. 2001. Coastal Zones and Marine Ecosystems. In *Climate Change 2001: Impacts, Adaptation and Vulnerability. Contribution of Working Group II to the Third Assessment Report of the Intergovernmental Panel on Climate Change.* [McCarthy, J.J., O.F. Canziani, N.A. Leary, D.J. Dokken, and K.S. White (eds.)]. Cambridge University Press, Cambridge, U.K. and New York, NY.
<www.grida.no/climate/ipcc_tar/wg2/index.htm>. Accessed August 31, 2009.

Christ, M., and M. Pavlick. 2006. *Watershed Based Plan for the Decker Creek Watershed, Preston and Monongalia Counties, West Virginia.* Friends of Deckers Creek, Dellslow, WV.

COEICC NRC (Committee on Ecological Impacts of Climate Change, National Research Council). 2008. *Ecological Impacts of Climate Change.* The National Academies Press, Washington, DC.

CTGR (Confederated Tribes of Grand Ronde). 2008a. *Confederated Tribes of Grand Ronde Nonpoint Source Assessment Report.* Confederated Tribes of Grand Ronde, Grande Ronde, OR.

CTGR (Confederated Tribes of Grand Ronde). 2008b. *Confederated Tribes of Grand Ronde Nonpoint Source Management Plan.* Confederated Tribes of Grand Ronde, Grande Ronde, OR.

IPCC (Intergovernmental Panel on Climate Change). 2007a. *Climate Change 2007: Impacts, Adaptation and Vulnerability. Contribution of Working Group II to the Fourth Assessment Report of the Intergovernmental Panel on Climate Change* [Parry, M.L., O.F. Canziani, J.P. Palutikof, P.J. van der Linden, and C.E. Hanson (eds.)]. Cambridge University Press, Cambridge, U.K.
<www.ipcc.ch/publications_and_data/publications_ipcc_fourth_assessment_report_ wg2_report_impacts_adaptation_and_vulnerability.htm>. Accessed August 31, 2009.

IPCC (Intergovernmental Panel on Climate Change). 2007b. *Climate Change 2007: Synthesis Report. Contribution of Working Groups I, II and III to the Fourth Assessment Report of the Intergovernmental Panel on Climate Change* [Pachauri, R.K, and A. Reisinger (eds.)]. IPCC, Geneva, Switzerland.
<www.ipcc.ch/publications_and_data/publications_ipcc_fourth_assessment_report_ synthesis_report.htm>. Accessed August 31, 2009.

IPCC (Intergovernmental Panel on Climate Change). 2008. *Climate Change and Water*. IPCC Technical Paper VI. <*www.ipcc.ch/pdf/technical-papers/climate-change-water-en.pdf*>. Accessed August 31, 2009.

JST (Jamestown S'Klallam Tribe). 2007. *Protecting and Restoring the Waters of the Dungeness: A Watershed-Based Plan Prepared in Compliance with Section 319 of the Clean Water Act*. Jamestown S'Klallam Tribe, Sequim, WA.

KT (Karuk Tribe), Department of Natural Resources. *Eco-Cultural Resources Management Plan*. Karuk Tribe, Happy Camp, CA.

PTOI (Puyallup Tribe of Indians). 2008. *Puyallup Tribe of Indians Nonpoint Management Plan*. Puyallup Tribe of Indians, Puyallup, WA

RCBLSC (Red Cliff Band of Lake Superior Chippewa). 2008a. *Tribal Nonpoint Source Assessment Report*. Red Cliff Band of Lake Superior Chippewa, Environmental Department, Red Cliff, WI.

RCBLSC (Red Cliff Band of Lake Superior Chippewa). 2008b. *Tribal Nonpoint Source Management Plan*. Red Cliff Band of Lake Superior Chippewa, Environmental Department, Red Cliff, WI.

RLBCI (Red Lake Band of Lake Superior Chippewa). 2008a. *Nonpoint Source Assessment of Streams, Lakes and Wetlands*. Red Lake Band of Chippewa Indians, Red Lake, MN.

RLBCI (Red Lake Band of Lake Superior Chippewa). 2008b. *Nonpoint Source Management Plan*. Red Lake Band of Chippewa Indians, Red Lake, MN.

SBLI (Soboba Band of Luiseño Indians). 2007. *Nonpoint Source Assessment Report and Management Program*. Soboba Band of Luiseño Indians, San Jacinto CA.

SMSC (Shakopee Mdewakanton Sioux Community). 2008. *Nonpoint Source Assessment Report*. Shakopee Mdewakanton Sioux Community, Prior Lake, MN.

ST (Suquamish Tribe). 2008. *Nonpoint Source Assessment and Management Program*. Suquamish Tribe, Suquamish, WA.

STOI (Stillaguamish Tribe of Indians). 2008. *Stillaguamish Tribe of Indians Nonpoint Source Management Plan*. Stillaguamish Tribe of Indians, Arlington, WA.

SYBCI (Santa Ynez Band of Chumash Indians). 2006a. *Non-point Source Assessment Report*. Santa Ynez Band of Chumash Indians, Santa Ynez, CA.

SYBCI (Santa Ynez Band of Chumash Indians). 2006b. *Non-point Source Management Plan.* Santa Ynez Band of Chumash Indians, Santa Ynez, CA.

UMUT (Ute Mountain Ute Tribe). 2005a. *Nonpoint Source Assessment for the Ute Mountain Ute Reservation of Colorado, New Mexico and Utah.* 2005 Revision. Ute Mountain Ute Tribe, Towaoc, CO.

UMUT (Ute Mountain Ute Tribe). 2005b. *Ute Mountain Ute Tribe Nonpoint Source Management Program Plan for the Ute Mountain Ute Reservation of Colorado, New Mexico and Utah.* 2005 Revision. Ute Mountain Ute Tribe, Towaoc, CO.

www.ingramcontent.com/pod-product-compliance
Lightning Source LLC
Chambersburg PA
CBHW080808180526
45168CB00006B/2365

United States Environmental Protection Agency
Office of Water
Nonpoint Source Control Branch–4503(T)
1200 Pennsylvania Avenue, NW
Washington, DC 20460

EPA 841-B-10-001

February 2010